ENVIRONMENTAL HEALTH - PHYSICAL, CHEMICAL AND BIOLOGICAL FACTORS

OIL SPILLS IN ARCTIC WATERS

AN INVENTORY OF RESEARCH EFFORTS

ENVIRONMENTAL HEALTH - PHYSICAL, CHEMICAL AND BIOLOGICAL FACTORS

Additional books in this series can be found on Nova's website under the Series tab.

Additional e-books in this series can be found on Nova's website under the e-book tab.

ENVIRONMENTAL HEALTH - PHYSICAL, CHEMICAL AND BIOLOGICAL FACTORS

OIL SPILLS IN ARCTIC WATERS

AN INVENTORY OF RESEARCH EFFORTS

NORA LINWOOD
EDITOR

nova publishers
New York

For permission to use material from this book please contact us:
Telephone 631-231-7269; Fax 631-231-8175
Web Site: http://www.novapublishers.com

Additional color graphics may be available in the e-book version of this book.

Library of Congress Cataloging-in-Publication Data

ISBN: 978-1-63321-580-1

Published by Nova Science Publishers, Inc. † New York

Contents

PREFACE

Chapter 1 – This white paper is a compilation of research on oil spills in ice-covered Arctic waters and recommendations for future work. Research entities in governmental, nongovernmental, industrial, and private organizations are identified. An introduction to the topic is provided.

Chapters 2 and 3 – This matrix is a selection of Arctic oil spill research drawn largely from US government sources and supplemented by representative projects from other sources including foreign governments and privately funded programs. The collection of references are limited to those available in the public domain.

Chapter 4 – In this white paper, the U.S. Arctic Research Commission (USARC) recommended to the federal government an invigorated oil spill research effort in the Arctic and a funding strategy that did not require new fiscal appropriation. Considering the potential for increased energy exploration and production in deeper, offshore waters, as well as an anticipated escalation in shipping in a rapidly evolving marine environment, the risks of oil spills remain a real and growing challenge. In developing these recommendations, USARC worked closely with the federal Interagency Coordinating Committee on Oil Pollution Research (ICCOPR), stakeholders, and the public.

Chapter 5 – Congress passed the Oil Pollution Act in 1990 (OPA). Among other things, OPA established the Interagency Coordinating Committee on Oil Pollution Research (interagency committee) to coordinate an oil pollution research program among federal agencies, including developing a plan, having the National Academy of Sciences review that plan, and reporting to Congress on the interagency committee's efforts biennially. The 2010 *Deepwater Horizon* explosion and fire led to the largest oil spill in U.S. history, raising new concerns about the effects of oil spills.

GAO was asked to assess the extent to which the interagency committee has facilitated the coordination of federal agencies' oil pollution research. (The Chairman, Subcommittee on Energy and Environment, House Committee on Science and Technology, now retired; and Representative Woolsey initiated this request.) In part, GAO analyzed committee documents and biennial reports and interviewed agency officials and nonfederal research entities.

In: Oil Spills in Arctic Waters
Editor: Nora Linwood
ISBN: 978-1-63321-580-1

Chapter 1

OIL SPILLS IN ARCTIC WATERS: AN INTRODUCTION AND INVENTORY OF RESEARCH ACTIVITIES AND USARC RECOMMENDATIONS*

United States Arctic Research Commission and US Army Corps of Engineers Cold Regions Research and Engineering Laboratory

Cover photo: Oil rig in the Beaufort Sea. Photo credit: Ocean Photography.

* This is an edited, reformatted and augmented version of a white paper issued November 2012.

Introduction

This "white paper" is a compilation of research on oil spills in ice-covered Arctic waters and recommendations for future work. We identify research entities in governmental, nongovernmental, industrial, and private organizations, and provide an inventory of research projects. Given that much work is currently in progress, we provide only a snapshot in time, and an introduction to the topic.

While an in-depth evaluation of the research results, a prioritization of research and development gaps, and a critical examination of the connection between research and oil spill response capability are undoubtedly important topics, they are beyond the scope of this effort. We hope that our foray into this subject encourages others to address these critical topics. Finally, the US Arctic Research Commission (USARC) closes this paper with recommendations for additional research on the topic of oil spills in ice-covered Arctic waters.

Motivation

We wrote this paper for several reasons. First, because interest in oil and gas development in the Arctic is on the rise, as is marine shipping, the likelihood of oil spills is increasing. Climate change, the retreat of Arctic ice, and global economic demand suggest that oil and gas prospects will be explored and eventually developed on the outer continental shelf of Alaska and in remote, icy waters of other Arctic nations. Increased Arctic marine transportation, and greater oil and gas exploration and production, amplify the possibility of oil spills.

Prior exploration and limited development have occurred in this region, but recent spills in other areas, such as the Gulf of Mexico and the North Sea, have heightened awareness and concern regarding the adequacy of spill prevention and cleanup techniques in ice-infested waters, including those characterized by intact sea ice sheets, broken ice conditions, and slushy conditions.

Second, we attempt to address the basic question, "What research is being done to prevent or respond to oil spills in the Arctic?" As it turns out, several entities are engaged in this activity, including domestic and foreign governments (federal and state), industry, nonprofit organizations, and private consultants.

The third reason is timeliness. Royal Dutch Shell has demonstrated a long-term interest in the outer continental shelf of Alaska, and had planned, in summer 2012, substantial Arctic exploratory drilling operations on two lease holdings in the Beaufort and Chukchi Seas. Because of several challenges, including damage to a spill containment dome (aka “capping stack”) during a testing accident, Shell decided, on September 17, 2012, to postpone deep (beyond 1400 feet) drilling in the Alaskan Arctic for at least a year.

Fourth, because oil spill preparedness is a nuanced issue, where the biases of supporters and detractors sometimes lead to selective highlighting, which may confuse the public, we try to present here an unbiased inventory of research, as well as some impartial recommendations on the topic.

Some praised Shell’s significant preparation efforts and safety plans, but others raised questions about the feasibility of oil spill response in remote regions. Spills in ice-infested waters and deployment of cleanup technologies pose many challenges. Advocates of offshore oil exploration and development in the Arctic point to the shallow-water locations of these low-pressure-formation prospects and the lack of problems during past drilling efforts. Beyond the obvious logistical and infrastructure challenges of Arctic exploration and development, opponents cite the paucity of techniques for effective cleanup of spills in and under ice, and the uncertain fate and transport of oil in the Arctic Ocean. Concerns have also been raised about the likelihood of long-term impacts on the environment, on the subsistence culture of Alaska Natives and other local inhabitants, and about their dependence on a clean and healthy ecosystem.

Given the continuing discussion about oil and gas development in the Arctic, it is important to review and attempt to answer this question, “What research is being done on oil spills in ice-covered waters?”

OUTLINE

The paper begins with a summary of federal efforts in three topic areas: (1) spill demarcation and mitigation, including containment and countermeasures; (2) oil spill response technologies for cleanup and recovery of oil, including affiliated data management tools; and (3) the fate of oil and its effects on the environment.

Following this summary is a discussion of the work of the Interagency Coordinating Committee on Oil Pollution Research (ICCOPR), whose mission is to coordinate oil pollution research among the federal agencies. The third section describes research efforts by other federal and state agencies, universities, and private entities. The final section highlights non federal activities of interest, and plans for additional research.

The white paper focuses primarily on research from the last decade, but it also references prior results, back to the 1980s, because current initiatives build largely on that body of work. USARC's recommendations for future research are included at the end of the paper.

SCOPE

Our task is to develop a list of organizations and entities that are conducting research on oil spills in Arctic waters and an inventory of research projects on that subject. We also provide a brief introduction to the research and the context within which it is conducted.

With regard to scope, we must also be clear as to what is not included in this white paper. Importantly, we stress that a scientific and technological assessment of the quality of the prior research results is beyond USARC's immediate abilities. Determining, in detail, whether research and development projects produced valuable results and measurable advances, or not, is certainly an important task, and one that should be undertaken by qualified researchers.

To this end, we note that the following text from the draft implementation strategy for National Ocean Policy (http:// www.whitehouse.gov/sites/default /files/microsites/ceq/national_ocean_policy_draft_implementation_plan_01- 12- 12.pdf) assigns such a duty to the US Coast Guard and the Bureau of Ocean Energy Management: "...assessments of current scientific research as well as traditional knowledge related to the impacts of resource development and pollution applicable to the Arctic.'

We also chose not to address the connection between research and the capacity to use research results in an actual oil spill response. And we do not assess the adequacy of in-place infrastructure, which is clearly critical to such a response. Although widely recognized as a critical and overarching concern, issues surrounding the regional response and recovery requirements, as well as systems and capacity, are topics we hope will be addressed by knowledgeable experts, perhaps motivated by this paper.

FUNDING

The allocation of funding for research is a fundamental issue. With respect to government funding for oil spill research, we rely on a report from the US Government Accountability Office (Report GAO-11-319, Federal Oil and Gas, March 2011, http://www.gao.gov/assets/320/317056.pdf) that cited a total cumulative expenditure of $164 million for the 11 years from 2000 to 2010 inclusive. Importantly, we readily acknowledge that some advances in oil spill response science, developed in non-Arctic regions, can certainly be applied to the Arctic.

Nevertheless, this expenditure ($164M) is significantly less than the funding dedicated to advance technology associated with oil exploration and production. It's also significantly less than what the government earns through lease sales. For example, compared to this amount, the 2008 Minerals Management Service (MMS) lease sale in the Chukchi Sea netted $2.7 billion, and Shell Oil is reported to have spent a total of $4.5 billion, thus far, in leasing costs and in preparation to explore their lease holdings in this area and in the Beaufort Sea.

In reviewing the manner in which the federal government funds research, USARC sees a significant shortcoming with respect to support for the National Oceanic and Atmospheric Administration (NOAA). We underscore the fact that NOAA is the only federal agency with oil spill preparedness, response, and restoration responsibilities under the Oil Pollution Act of 1990 (OPA90) that does not receive a direct appropriation from the Oil Spill Liability Trust Fund (OSLTF) for research and development. USARC strongly recommends that NOAA receive funding from the OSLTF.

Research funding should be provided for both basic research (commonly academic) and applied. While basic research has led to greater understanding, many other improvements in oil spill prevention and response have resulted from actual testing (both in the field and in test facilities), demonstrations and experimental trials, drills, literature and technology reviews, and hands-on application and experience.

Funding should be directed to projects that show the greatest promise for results that will improve spill prevention and response. Basic research should demonstrate excellent intellectual merit, while applied research should also be rigorously vetted and should be performance-based.

In short, USARC calls for greater funding for research to prevent and respond to oil spills in the Arctic. As George Orwell said, "We have now sunk to a depth at which restatement of the obvious is the first duty of intelligent men.'

US Federal Research

Arctic oil spill research conducted by the US federal government addresses environmental stewardship, marine transportation, and commercial development of natural resources. A small percentage of the revenue that the US government collects from the oil industry, through the Department of the Interior's (DOI) Office of Natural Resources Revenue (ONRR), is invested in research.

For example, the Bureau of Ocean Energy Management (DOI/BOEM) conducts research on the environmental and economic impact of developing the resource, and the Bureau of Safety and Environmental Enforcement (DOI/BSEE) focuses on oil spill prevention, preparedness, detection, and cleanup. Other examples are described below, and in an online table that accompanies this white paper (http://www. arctic.gov/publications/oil_spills_tableA.pdf). Government-funded research initiatives are primarily solicited via the Broad Agency Announcement (BAA) process, but other mechanisms, including research conducted within agencies, are used as well.

Industry Research

The oil and gas industry also conducts research on oil spill prevention and response in Arctic waters. A recent example is noteworthy. In January 2012, the International Association of Oil & Gas Producers (OGP) announced an Oil Spill Response Technology Joint Industry Program (JIP) focusing on oil spill recovery in ice and spill response in Arctic marine waters. Below, we discuss this JIP in detail.

One of the first tasks of this JIP has been to collect and review the current state of knowledge on the subject. Some of the relevant reports they identified include:

- American Petroleum Institute and the Joint Industry Program on Oil Spill Recovery in Ice, *Spill Response in the Arctic Offshore*, 2012, http://www.api.org/~/media/Files/EHS/Clean_Water/Oil_Spill_Prevention/Spill-Responsein-the-Arctic-Offshore.ashx

- SINTEF, Report no. 32, *Joint Industry Program on Oil Spill Contingency for Arctic and Ice-Covered Waters, Summary Report*, 2010, http://www.sintef.no/project/JIP_Oil_In_Ice/Dokumenter /publications/JIP-rep-no-32-Summaryreport.pdf
- Environmental Studies Research Funds, Canada, Report No. 177, *Beaufort Sea Oil Spills State of Knowledge Review and Identification of Key Issues*, 2010, http://www.esrfunds. org/pdf/177.pdf

TEST FACILITIES

US test facilities support research on the fate of oil in ice-covered waters. Two examples are the Ohmsett Facility in Leonardo, New Jersey, and the refrigerated facilities at the US Army Corps of Engineers Cold Region Research and Engineering Laboratory (CRREL) in Hanover, New Hampshire.

Ohmsett provides independent and objective performance testing of full-scale oil spill response equipment and marine renewable energy systems (wave energy conversion devices), and improves technologies through research and development. "OHMSETT" (http://www.ohmsett.com) was originally an acronym for Oil and Hazardous Materials Simulated Environmental Test Tank. As oil is the only hazardous material now tested, the name is now just "Ohmsett," and it is referred to as "The National Oil Spill Response Research & Renewable Energy Test Facility." It is owned by the US Navy, and is maintained and operated by BSEE through a contract with MAR Inc. The facility conducts tests with oil under various simulated sea states.

CRREL, which has expertise in cold regions science and technology, has refrigerated facilities to evaluate oil behavior in and under ice. Tests conducted at CRREL are summarized at http://www.crrel.usace.army.mil/innovations /oil_spill_research. The proximity of the two research facilities enables collaboration on projects, leveraging the expertise of research staff.

Currently, Alaska Clean Seas, a not-for-profit oil spill response cooperative whose membership includes oil and pipeline companies operating on the North Slope of Alaska, is collaborating with CRREL to design an oil and ice testing tank at Prudhoe Bay, Alaska. Two of the design criteria are: nonferrous construction to minimize electromagnetic interference for testing oil detection antennae, and adequate depth to test remotely and autonomously operated vehicles (ROV/AOV).

Importantly, these facilities are critically valuable research facilities because it is currently illegal in the United States to conduct in situ field testing of procedures to respond to oil spills by intentionally spilling oil in the natural environment.

INVENTORY OF RESEARCH PROJECTS ON OIL SPILLS IN ARCTIC WATERS

An inventory of 203 research projects on oil spills in Arctic waters has been compiled, and are listed in an online table that accompanies this white paper (http://www.arctic.gov/ publications/oil_spills_tableA.pdf). The projects, mostly funded by the US government, industry, and Canada, are categorized as follows:

a) Assessments and workshops (18 projects)
b) Fate and behavior of oil (5)
c) Environmental effects (3)
d) Detection, mapping and tracking (33)
e) Mitigation (116)
f) Testing standards (8)
g) Recent field demonstrations (post 2000) (9)
h) Miscellaneous (11)

For each project, we provide a description that includes the subdiscipline of research, relevant website links, a description of the project, the name of the organization that conducted the research, the name of the sponsor, the year of the study, the objective, and the name of the research facility (e.g., Ohmsett), if one was used.

Also online, courtesy of Mr. Larry Dietrick, Director of the Division of Spill Prevention and Response in the State of Alaska's Department of Environmental Conservation, is a table of the 45 project titles and descriptions for initiatives that have been supported by State of Alaska between the years 2000 and 2010 (http://www.arctic.gov/publications/ oil_spills_tableB.pdf). Note that these projects include Arctic and non-Arctic initiatives, as well as terrestrial and marine environments.

PRIOR RESEARCH PROVIDES A FOUNDATION

After consulting a variety of sources, including discussions with partners, reading the peer-reviewed and "gray" literature, tracking requests for proposals, and attending conferences, it is clear that a considerable amount of work has been done, and continues to be, and in a variety of research areas, to understand the behavior and fate of oil in ice-covered waters. This understanding provides a credible foundation for applied research and engineering designs aimed at minimizing the risk of an accident, and, should one occur, developing more effective response and recovery techniques.

The breadth and value of this body of knowledge may be underappreciated by the general public, and if so, greater efforts need to be made to communicate the value of this information to the layperson and to media representatives.

Nonetheless, three significant areas deserve additional research: preparedness, response effectiveness, and damage assessment.

Preparedness

- Developing oil detection and mapping techniques that are independent of ice conditions, rapidly deployable, and can be quickly interpreted in the field. Accurate assessment of the size and area of the oil spill will dictate the appropriate magnitude of the response effort. Timely assessment will minimize the impacted area. Techniques that need enhancement are satellite and airborne sensors for cursory assessment, and refined delineation via airborne and underwater systems.
- Establishing baseline environmental conditions and assessing environmentally sensitive areas. In studies commissioned by BOEM and by various nongovernmental organizations, much of the Chukchi and Beaufort coasts have been identified as areas of heightened ecological significance. Large swaths of the shoreline zones are typically shallow and composed of ice-rich permafrost soils (though experiencing thawing in many locations). The impacts of crude and dispersed oil deposition on these shorelines and the cleanup techniques available should be assessed.
- Determining the practicality of using robots and ROVs/AOVs to assist in the cleanup process. Such technology could improve the response

effectiveness in severe weather conditions, dark, dirty, or dangerous situations, or in unstable ice conditions.

- Field testing and training of techniques by the responders to assure familiarity with processes before they have to be deployed for a spill.
- Modeling of oil in, under, and within ice-infested waters. Better determination of how oil accumulates, disperses, and degrades (both biologically and nonbiologically).
- Researching options to minimize and/or mitigate the risk associated with vessel use and carriage of heavy-grade fuel oil in the Arctic.
- Compiling and integrating the datasets necessary for Arctic oil spill planning, coordination, and response that would be used in NOAA's Arctic Environmental Response Management Application (ERMA®), described below.
- Completing scientifically based field experiments and tests of response tools in US Arctic waters, if permits can be secured. If not, continue to conduct experiments in test tank facilities and partner with non-US entities in Norway and Canada to conduct field experiments in foreign waters.
- Identifying resource and infrastructure shortfalls for high-risk scenarios, and assessing strategies to address those shortfalls.
- Educating and involving stakeholders and the public on the various detection and mitigation techniques and baseline and subsistence issues.

Response Effectiveness

- Improving the performance of mechanical oil recovery systems in ice-infested water.
- Improving in situ burning techniques, which may minimize the volume of oil in the water.
- Quantifying the environmental effects of in situ burning, including smoke and residue products, so they may be included in the response assessment analysis, to assist in deciding the best approach for remediation as a function of in situ conditions.
- Improving the effectiveness of chemical herders that have been effective in scavenging oil from broken ice, expanding the window of opportunity for recovery, and improving the cleanup process.

Additional research may improve the performance of herders in Arctic conditions, characterized by ice, waves, and cold water.

- Evaluating dispersants in Arctic conditions. Dispersants were used extensively in the Deepwater Horizon oil spill cleanup. The effectiveness and toxicity of the dispersants, and their ability to disperse oil when injected at the source of submerged leaks, are topics of considerable research, but not under conditions that exist in polar regions.
- Evaluating the gaps in informational, logistical, communications, and infrastructure needs that currently exist in responding to spills, such as those described in the March 2008 report of the NOAA/UNH co-led Coastal Response Research Center (CRRC; http://www.crrc.unh.edu/ workshops/arctic_spill_summit/arctic_summit_report_ final.pdf).

Damage Assessment

- Undertaking research into the human dimension of spills. What are the impacts of Arctic spills on human beings? Factors to consider include the health and well-being of first responders and of local residents exposed directly and indirectly to petroleum products (such as through bioaccumulation and subsistence consumption of wildlife).
- Determining the effects of dispersed oil on the Arctic ecosystems, with assessments of the toxicity of dispersed oil and dispersants on benthic flora and fauna, and, in particular, marine mammals and sea birds.
- Analyzing the research and data needs for the Natural Resource Damage Assessment and Restoration Program (NRDA Restoration Program) process and for restoration projects.

Funding of Oil Spill Research

According to a recent report from the Government Accountability Office (GAO-11-319, March 2011), federal agencies that are members of ICCOPR have expended about $164M on oil pollution research from fiscal 2000 through 2010 (Table 1). Most of the funds ($145M) come from the Oil Spill Liability Trust Fund (OSLTF) authorized by the Oil Pollution Act of 1990

(OPA90). OSLTF is funded by a tax on petroleum production and importation, originally established at $0.05 per barrel, and later increased to $0.08 in 2008.

Beyond the trust funds, agencies have spent only an additional $18M on oil pollution research in over a decade, and only a fraction of that on spills in ice-covered waters, an area that ICCOPR has identified as an important need. Clearly, since the Deepwater Horizon incident, expenditures have increased, and the expenditures in 2011 and 2012 are likely higher than in recent years.

We note that NOAA is only federal agency with responsibilities for oil spill preparedness, response, and restoration, assigned by OPA90, which does not receive a direct appropriation from the OSLTF for research and development.

We further note that the manner in which research funds are expended by agencies varies widely. Some agencies, seeking expertise from beyond their own staff, issue calls for proposals and Broad Agency Announcements, and may also use a peer-review process to select projects for funding. Others follow a contractual approach, or expend the funds internally, within the agency.

With respect to the funding of the Oil Spill Recovery Institute (OSRI), created by OPA90, and described below, resources for oil spill research are tied to the interest earned on the *Exxon Valdez* oil spill settlement. Recently, funding for research has been approximately $1M per year.

Federal Arctic Marine Oil Spill Research Efforts

Department of the Interior

- Minerals Management Service (MMS)
- Bureau of Ocean Energy Management, Regulation and Enforcement (BOEMRE)
- Office of Natural Resources Revenue (ONRR)
- Bureau of Ocean Energy Management (BOEM)
- Bureau of Safety and Environmental Enforcement (BSEE)
- US Geological Survey (USGS)

Table 1. Federal agency funding for oil pollution research, both trust and agency funds, FY 2000–2010 in $M

Agency	2000	2001	2002	2003	2004	2005	2006	2007	2008	2009	2010	Total
BOEMRE*	$7.1	$6.9	$6.8	$6.7	$7.5	$7.9	$7.5	$7.5	$7.1	$6.6	$6.2	$77.8
USCG	$4.7	$4.8	$4.8	$4.3	$4.1	$2.3	$2.8	$2.1	$2.1	$0.6	$0.7	$33.3
EPA	$1.1	$1.1	$1.1	$1.1	$1.0	$1.0	$0.9	$0.8	$0.9	$0.7	$0.6	$10.3
NASA	$0.0	$0.0	$0.0	$0.0	$0.0	$0.0	$0.0	$0.0	$0.4	$0.0	$0.0	$0.4
Navy	$4.3	$3.5	$1.2	$0.4	$0.3	$0.3	$0.2	$0.3	$0.6	$0.4	$0.6	$12.1
NOAA	$0.0	$0.0	$0.0	$0.0	$2.3	$2.2	$3.3	$3.2	$0.0	$0.0	$0.0	$11.0
PHMSA**	$0.0	$0.0	$0.0	$0.0	$3.4	$3.3	$2.1	$2.0	$3.4	$2.2	$2.2	$18.6
Annual Total	$17.2	$16.3	$13.9	$12.5	$18.6	$17.0	$16.8	$15.9	$14.5	$10.5	$10.3	$163.5

*BOEMRE was formerly MMS. Expenditures include about $3M/yr for the operation and maintenance of the Ohmsett facility.

** PHMSA is Pipeline and Hazardous Materials Safety Administration.

Data from the March 2011 GAO-11-319 report on federal oil and gas titled, *Interagency Committee Needs to Better Coordinate on Oil Pollution Prevention and Response*, http://www.gao.gov/assets/320/317056.pdf.

MMS, BOEMRE, ONRR

Research programs funded by MMS, BOEMRE, BOEM, and BSEE are listed in an accompanying online table (http://www. arctic.gov/publications /oil_spills_tableA.pdf).

The Minerals Management Service (MMS) was created on January 19, 1982, nearly 30 years after Congress passed two milestone Acts—the Submerged Lands Act and the Outer Continental Shelf (OCS) Lands Act. This date was also 28 years after the first OCS lease sale was held. MMS was charged with resource management, safety, environmental protection, and revenue collection.

Most of the reports that analyzed the Deepwater Horizon disaster issued recommendations for industry and government, including changes in MMS structure and procedures. MMS had a broad mandate with inherently conflicting missions, and was under-resourced for the remarkable expansion of offshore drilling that took place over the last several decades.

On May 19, 2010, Department of Interior Secretary Salazar signed a Secretarial Order that separated MMS into three entities, the Bureau of Ocean Energy Management (BOEM), the Bureau of Safety and Environmental Enforcement (BSEE), and the Office of Natural Resources Revenue (ONRR). These entities oversee energy leasing, safety and environmental protection enforcement, and revenue collection, respectively. On June 21, 2010, MMS was renamed Bureau of Ocean Energy Management, Regulation, and Enforcement (BOEMRE). To complete the reorganization, BOEMRE was replaced by BOEM and BSEE on October 1, 2011.

MMS/BOEMRE has historically leveraged research funds with industry and others to conduct research and development of oil spill technologies in Arctic waters. From 1997 through 2008, MMS/BOEMRE successfully developed, implemented, and conducted 31 projects directly related to Arctic oil spill response. More than 40% of these projects were jointly funded by state and federal government agencies, academia, private industry, and foreign governments.

The Oil Spill Response Research Program, described in the 2009 report *Arctic Oil Spill Response Research and Development Program: A Decade of Achievement* (http://www.iccopr.uscg.gov/iccopr/i/files/MMSArctic Research _2009.pdf) focused on remote sensing and surveillance, mechanical response, chemical treating agents including dispersants, and in situ burning. This research included conducting operational research experiments with oil in small and large test tanks, at sea, and in ice to test many different types of oil spill response technologies and methodologies.

A review of this paper by the World Wildlife Fund, published in December 2009 and prepared with the assistance and technical expertise from Harvey Consulting, LLC, is titled, *Not So Fast: Some Progress in Spill Response, but US Still Ill-Prepared for Arctic Offshore Development.* It focuses on six findings regarding oil spill response in Arctic waters, and concludes, "...despite progress, significant gaps remain in the availability of effective oil spill response tools for the Arctic." WWF also called for "a response gap analysis in the Alaskan Arctic that would fully disclose and quantify the percentage of time during which local conditions exceed the demonstrated limits of spill response systems."

We note that the Canadian federal government has conducted such a response gap analysis for the Canadian Beaufort Sea and the Canadian Davis Strait. Canada's National Energy Board commissioned SL Ross Environmental Research for the work, and the report is available at https:// www.neb-one.gc.ca/ll-eng/livelink.exe?func=ll&objId=702787&objAction= browse.

The Environmental Studies Program (http://www.boem. gov/Studies) has a broad remit to inform policy decisions on the development of the OCS by developing, conducting, and overseeing scientific research efforts. By region, Alaska receives significant attention, and research focuses on protected and endangered species, oceanography, biology, subsistence and traditional knowledge studies, and economic modeling. Other studies consider the fate and effects of oil in ice, and remote sensing and modeling of physical and biological systems. These programs have now been distributed between BOEM and BSEE, as described below.

BOEM

The Bureau of Ocean Energy Management (BOEM) manages the exploration and development of the nation's offshore resources. It seeks to appropriately balance economic development, energy independence, and environmental protection through oil and gas leases, renewable energy development, and environmental reviews and studies. BOEM is responsible for the Five Year OCS Oil and Gas Leasing Program that includes assessments, inventories, and production projections. BOEM administers oil and gas lease sales and offshore renewable energy programs, and conducts environmental reviews and studies. Of particular interest to this white paper are the Alaska Annual Study Plans and the Alaska Region Ongoing Studies that divide efforts into categories of Physical Oceanography, Fates and Effects, Habitat and Ecology, Marine Mammals and Protected Species, Social

Systems, Information Management, and Integrated Studies. The latest compilation of ongoing studies is on the BOEM website at http://www.boem.gov/uploadedFiles/BOEM/Environmental_Stewardship/Environmental_Studies/Alaska_Region/Alaska_Studies/Alaska_Ongoing_Studies.pdf.

In addition to current studies on Arctic biological communities (e.g., fish, sea birds, benthos, and marine mammals), the research program includes modeling of Arctic Ocean meteorology, circulation patterns, currents and their tracking by high-frequency radar, and the location and recurrence of polynyas (areas of open seawater surrounded by sea ice).

The University of Alaska Coastal Marine Institute (CMI) was created by a cooperative agreement between the University of Alaska and the US Department of the Interior Minerals Management Service Alaska Region to study coastal topics associated with the development of natural gas, oil, and minerals in Alaska's outer continental shelf. Many of the CMIfunded projects address issues related to fisheries, biomonitoring, physical oceanography, and the fate of oil. Research projects funded by the CMI are required to have at least one investigator from the University of Alaska. Cooperative research between University of Alaska scientists and various state agencies is also encouraged. Current research topics include movement and tracking of sea ice and polynyas, fish and benthic organism assessments, subsistence use, and effects of climate change. A full description of the CMI and their activities can be found at http://www.sfos.uaf.edu/cmi.

BSEE

The mission of the Bureau of Safety and Environmental Enforcement (BSEE) is to promote safety, protect the environment, and conserve resources offshore. To ensure that industry operations on the OCS incorporate the best available and safest technologies, BSEE attempts to enhance response technologies and capabilities through the Oil Spill Response Research (OSRR) Program. For more than 25 years, OSRR has funded a comprehensive and enduring research program to improve oil spill response technologies. The major focus of the program is to improve the knowledge and technologies used to detect, contain, and clean up oil spills on the US OCS. Current OSRR projects cover a wide spectrum of oil spill response issues and include laboratory, mesoscale, and full-scale field experiments. Major topic areas, some of which are studied at Ohmsett, include remote sensing and detection, physical and chemical properties of crude oil, mechanical containment and recovery, chemical treating agents and dispersants, in situ burning, and deepwater operations.

The Ohmsett oil spill research facility, located in Leonardo, New Jersey, is operated and maintained by BSEE through a contract with MAR Inc. of Rockville, Maryland. The facility is the largest outdoor saltwater wave/tow tank facility in North America and is the only facility where full-scale oil spill response equipment testing, research, and training can be conducted in a marine environment with oil under controlled environmental conditions (waves and oil types). The tank's wave generator creates realistic sea environments, while state-of-the-art data collection and video systems record test results. The tank has been used to test oil spill containment/ cleanup equipment and techniques, to test new designs in response equipment, to validate research findings, and to conduct training with actual oil spill response technologies. Partnering with CRREL, oil spills mitigation tests in ice-infested water (created using ice blocks) have been conducted at the Ohmsett facility. These tests include the Mechanical Oil Recovery in Ice-Infested Waters (MORICE) system and testing of dispersants in conjunction with a wave field. For more information see http://ohmsett.com/facility.html.

BSEE also operates the National Offshore Training Center with a curriculum that is designed to train new inspectors and to keep others up to date on emerging technologies and processes.

In December 2011, BSEE posted Broad Agency Announcement Number E12PS00012 for Proposed Research on Oil Spill Response Operations in the US Outer Continental Shelf. Two of the 10 topics in the BAA focused on ice and oil.

Additional helpful urls include:

- BSEE's website for oil spill response research: http://www.bsee.gov/Research-and-Training/Oil-Spill- Response-Research-(OSRR).aspx
- BSEE's Arctic oil spill response projects: http:// www.bsee.gov/Research-and-Training/Technology-Assessment-and-Research/tarprojectcategories/ArcticOil-Spill-Response-Research.aspx
- BSEE's two most recent (FY12) projects: http://www.bsee. gov/Research-and-Training/Technology-Assessmentand-Research/Project1007.aspx and http://www.bsee.gov/ Research-and-Training/Technology-Assessment-andResearch/Project1000.aspx

Oil herding test conducted by I. Buist of SL Ross Project 617. Photo courtesy of CRREL.

USGS

In order to pursue a comprehensive, science-based approach to energy development on the OCS, Department of Interior Secretary Salazar directed the US Geological Survey (USGS) to determine the gaps in scientific understanding of OCS energy development in the Arctic, particularly in the Chukchi and Beaufort Seas. That report was released on June 23, 2011, and is available at http://pubs.usgs.gov/circ/1370/pdf/ circ1370.pdf. The report summarizes the large volume of existing scientific information, identifies where knowledge gaps exist, and provides initial guidance on new and continuing research that could improve decision-making.

Among the major areas noted in the report where additional scientific research, analysis, and synthesis could reduce uncertainties include the following:

- Developing a better understanding of the effects of climate change on physical, biological and social conditions as well as resource management strategies in the Arctic
- Developing foundational geospatial data on the Arctic Outer Continental Shelf
- Synthesizing existing scientific information on a wide range of topics on the Arctic

- Building upon advances in spill-risk evaluation and response knowledge by developing better information on key inputs to spill models (such as oceanographic, weather, and ecological data)
- Improving dialogue and using collaborative, comprehensive science planning, both domestically and internationally

A fact sheet on the Arctic study is available at http://pubs.usgs.gov/fs/2011/3048. The full report is available at http://pubs.usgs.gov/circ/1370.

Department of Commerce

- National Oceanic and Atmospheric Administration (NOAA)

The mission of the National Oceanic and Atmospheric Administration's Office of Response and Restoration (OR&R) is to prepare for, evaluate, and respond to threats to coastal environments, including oil and chemical spills, releases from hazardous waste sites, and marine debris. To address the increased potential for oil spills in the Arctic due to increased vessel traffic and oil exploration and development, NOAA OR&R has several initiatives currently underway. For further information see http://response.restoration.noaa.gov/oil-and-chemical-spills/oil-spills/activities- arctic.html.

- The Arctic Environmental Response Management Application (ERMA®) is a web-based GIS tool that assists both emergency responders and environmental resource managers in addressing incidents that may harm the environment. ERMA® integrates and synthesizes real-time and static data (information such as the extent and concentration of sea ice, locations of ports and pipelines, and vulnerable environmental resources) into a single interactive map, providing a quick visualization of the situation and improving communication and coordination among responders and environmental stakeholders. The Arctic ERMA® was developed and supported through a partnership between the NOAA Office of Ocean and Coastal Resource Management, the Oil Spill Recovery Institute, BSEE, and the University of New Hampshire. Further information can be found at: http://response.restoration.noaa. gov/maps-and-spatial-data/environmental-responsemanagement-application-erma/arctic-erma.html.

- NOAA OR&R is the lead science adviser to the US Coast Guard during oil spill response operations in Alaska and it maintains a role on the Alaska Regional Response Team. OR&R's expertise spans oceanography, biology, chemistry, and geology, allowing the response team to estimate oil and chemical trajectories, analyze chemical hazards, and assess risks to coastal animals, habitats, and important areas to humans. This team is led by regional Scientific Support Coordinators. OR&R has also initiated the Arctic Joint Assessment Team and has assisted the Arctic Council, Canada, and Norway in developing Arctic spill response techniques and plans for spills in Arctic waters.
- Baseline ecosystem information is also being collected to properly identify environmental risks as well as to determine the current status and health of Arctic natural resources for NRDA Restoration Program purposes.

NOAA OR&R has also developed and funded a cooperative research partnership with the University of New Hampshire (UNH) to further oil spill research. The NOAA/UNH co-led Coastal Response Research Center (CRRC) was established in 2004. The CRRC has supported many research projects, workshops, and working groups on the subject of oil spill research and development, and has published a variety of reports. The Center for Spills in the Environment (CSE) is another center, also at UNH, that expands the scope of interaction and cooperation with the private sector, other government agencies and universities. Both the CRRC and the CSE are administered by, and located at, the UNH campus in Durham, New Hampshire.

The CRRC participated in the Selskapet for INdustriell og TEknisk Forskning ved norges tekniske hoegskole (SINTEF; the Foundation for Scientific and Industrial Research at the Norwegian Institute of Technology) Joint Industry Program "Oil in Ice" through research efforts associated with detection of dissolved polycyclic aromatic hydrocarbons (PAHs) from oil spills using passive samplers in cold water and ice cores; transport, fate, and potential exposure of oil in ice; oil movement in sea ice; and biodegradation of oil in Arctic first-year sea ice. Current studies being supported by the CRRC and NOAA are investigating the effectiveness, physical transport, chemistry, and biological effects of chemical dispersants, in particular, those used during the Deepwater Horizon incident. Other activities include conducting studies on the physical and chemical changes in heavy oil as well as holding workshops on oil spill response issues. Workshops focusing solely on the Arctic include:

- Northwest Arctic Borough: An Oil Spill Workshop (May 22–23, 2012)
- Arctic ERMA® (April 5–6, 2011)
- NRDA in Arctic Waters: The Dialogue Begins (April 20–22, 2010)
- US Coast Guard Arctic Response (April 23, 2010)
- Opening the Arctic Seas: Envisioning Disasters and Framing Solutions (March 18–20, 2008)

A full description of CRRC and CSE activities can be found at http://www.crrc.unh.edu.

Department of Defense (DoD)

- US Army Corps of Engineers (USACE)
- US Navy (USN)

Because DoD mission priorities are currently focused on regions outside the Arctic, DoD is not providing any direct funds for research related to oil spills in ice-covered waters. DoD also acknowledges that, consistent with the current and projected level of DoD activity in the Arctic, it is most likely that an oil spill event will occur within the civilian sector. Under these circumstances, DoD will play a supporting role in responding to the spill, as it did for the Deepwater Horizon response. While not providing direct research support, DoD is making important contributions on other levels, including:

a) Maintaining access to facilities and expertise at the US Army Corps of Engineers' Cold Regions Research and Engineering Laboratory (CRREL) to conduct research on all aspects oil spills in ice-covered waters. CRREL is an international center of expertise on cold regions science and technology with cold rooms and refrigerated hydraulic facilities. Facilities used for oil and ice research include the Ice Engineering Facility (IEF), Geophysical Research Facility (GRF), and the Materiel Evaluation Facility (MEF). While CRREL does not directly fund these studies, their unique facilities, and in-house experts provide the capabilities to test equipment and techniques in near-field conditions. Staff at CRREL and Ohmsett have partnered on research projects to capitalize on the synergy of the researchers at DoD and DOI. Oil spill projects conducted at CRREL are summarized at http://www.crrel.usace.army. mil/innovations/oil_spill_research.

b) Scenario-driven exercises to evaluate DoD's level of readiness for Arctic operations, including communication protocols and logistic and equipment support. Knowledge and technology gaps are routinely identified during these exercises and reported, highlighting opportunities for future work for DoD and non-DoD entities alike. Recently, these exercises have included oil spill scenarios that are sufficiently severe to require a request of support from the DoD.
   - US Navy Fleet Arctic Operations Game, US Naval War College, Newport, RI, September 13–16, 2011
   - NORAD/USNORTHCOMM Arctic Collaborative Workshop, National Defense University, Washington, DC, March 13–15, 2012

c) The Emergency Ship Salvage Material (ESSM) System consists of facilities that provide salvage and pollution control (Oil Spill Response [OSR]) equipment and operators on an emergency basis. The worldwide ESSM facilities routinely provide mutual augmentation of personnel and equipment. One such facility is located on Joint Base Elmendorf Richardson (JBER). The ESSM OSR equipment is designed to be transportable and sustainable in the field. Although not specifically designed for Arctic use, the equipment has been modified for the Arctic environment but not for sea ice. Lake ice, in shore, and clear sea operations are all exercised annually. Systems located at JBER include two Vessel of Opportunity Skimming Systems, three 36-foot self-propelled belt skimming systems, boom, bladders, boats, inshore skimmers, support vans, and lightering systems.

Department of Homeland Security (DHS)

- US Coast Guard (USCG)

The United States Coast Guard operates the Research and Development Center (RDC), located in New London, Connecticut. This center is Coast Guard's only facility that conducts research, development, testing, and evaluation in support of their missions. The center evaluates the feasibility and affordability of mission execution solutions and provides operational and risk-management analysis for all stages of the acquisition process. When appropriate, the RDC collaborates with relevant professionals in the public and private sectors.

The Research, Development, Test, and Evaluation (RDT&E) Program conducts applied scientific research, development, testing, and evaluation of new technologies. The program pursues technologies that provide incremental improvements as well as those with the greatest potential to strategically transform the way the Coast Guard functions. As part of the program, the RDC is evaluating new technologies and techniques for cold weather oil recovery and is documenting performance gaps. A series of increasingly complex demonstrations have been started in the Great Lakes, and are also planned in Alaska. The RDC just completed the second oil-inice demonstration within Sector Sault Ste. Marie at St. Ignace, Michigan. The objective of this effort was to evaluate response capabilities in cold weather by leveraging Coast Guard and other local assets before conducting a more complex demonstration in Alaska. The demonstration scenario involved a fabricated leak from a pipeline in the open waters of the Mackinac Straits in northern Michigan. Tests were conducted using a helix skimmer in ice and also fire booms to separate out sections of the ice cover. Other R&D projects include sensors to detect submerged oil spills by multibeam sonar, a laser line scan system, fluorescent spectroscopy, and in situ mass spectrometers.

The USCG held a two-day workshop on "Leadership for the Arctic" in April 2012 to discuss science and research, maritime safety, maritime stewardship, legal issues, and governance. This meeting helped develop context for Arctic policy decisions as the USCG prepared for the planned Royal Dutch Shell exploration activities in the Beaufort and Chukchi Seas in the summer of 2012, via the exercise Arctic Shield 2012.

Another joint tabletop exercise was held by Shell Oil on May 24, 2012, to exercise their Chukchi Sea oil spill response plan. In addition to the USCG, participants included representatives from DOI, EPA, NOAA, and the State of Alaska.

Throughout the summer of 2012, USCG cutters and aircraft conducted exercises and training off Alaska to implement plans to mitigate potential impacts from a pollution incident. At the same time, scientists at the USCG's RDC were developing new ways to rapidly deploy and support response equipment at an incident scene. Demonstrations of equipment and tactics were held on the Great Lakes, and additional exercises are planned for the winter of 2013.

Other Federal Agencies

Other federal agencies, including the Environmental Protection Agency (EPA), the National Institute of Standards and Technology (NIST), the Department of Energy (DOE), the Department of Interior's Fish and Wildlife Services (FWS), and the National Aeronautics and Space Administration (NASA) conduct research on oil spills, but because their research is not specifically in association with ice-covered waters, they are not described in detail in this report.

The interagency "Committee on Marine Transportation Systems" has created a "US Arctic Marine Transportation Integrated Action Team" (http://www.cmts.gov/Activities/ ActionTeams.aspx) led by USCG, the US Department of Transportation Maritime Administration (MARAD), and NOAA. It is focusing on issues related to marine transportation in the Arctic. While not focusing on basic research, the group has discussed issues associated with applied research.

COORDINATION OF FEDERAL R&D EFFORTS

Many entities are involved in the funding and conduct of oil spill research. With declining federal budgets, it is important to avoid duplication and effectively establish the means to coordinate activities and joint funding opportunities.

The Interagency Coordinating Committee on Oil Pollution Research (ICCOPR) was established under to Section 7001(a) of the Oil Pollution Act of 1990 (PL 101-380, also referred to as OPA90 in this report). Chartered as a 13-member interagency committee, its mission is to "coordinate a comprehensive program of oil pollution research, technology development, and demonstration among the federal agencies, in cooperation and coordination with industry, universities, research institutions, state governments, and other nations, as appropriate, and shall foster cost-effective research mechanisms, including the joint funding of the research."

One of the tools that the ICCOPR uses to communicate their activities is the ICCOPR Biennial Report to Congress. This biennial report, required by

Section 7001(e) of the Oil Pollution Act of 1990, describes the activities carried out by the ICCOPR in the preceding two fiscal years as well as activities proposed to be carried out in the next two fiscal years. All of the ICCOPR biennial reports from 1992 to present day are available on the ICCOPR public website with the exception of the report from 1999–2000, which was not submitted due to the Federal Reports Elimination and Sunset Act of 1995.

In addition to producing biennial reports, ICCOPR also maintains an Oil Pollution Research and Technology Plan. Development of an Oil Pollution Research and Technology Plan is required by Section 7001(b) of the Oil Pollution Act of 1990. ICCOPR prepared the original Oil Pollution Research and Technology Plan to define the roles of each federal agency involved in oil spill research and development. The original plan was submitted to Congress in April 1992. The National Research Council's Committee on Oil Spill Research and Development, under the auspices of the Marine Board, reviewed this plan. Using input from the Marine Board, ICCOPR revised the plan in May 1993 to address spill prevention, human factors, and the field testing/demonstration of developed response technologies. The current version of the plan, still based on Marine Board recommendations, is dated April 1997. ICCOPR is updating the 1997 Oil Pollution Research and Technology Plan and expects to be finished in Fiscal Year 2013. Once completed, this plan will be available to provide oil pollution research guidance to government, industry, and academia.

The US Coast Guard chairs ICCOPR, and the current membership (now standing at 14 since the split of MMS/BOEMRE into BOEM and BSEE) includes:

- Department of Commerce—National Oceanic and Atmospheric Administration*
- Department of Commerce—National Institute of Standards and Technology
- Department of Energy
- Department of the Interior—US Fish & Wildlife Service
- Department of the Interior—Bureau of Safety and Environmental Enforcement*
- Department of the Interior—Bureau of Ocean Energy Management*
- Department of Transportation—Maritime Administration
- Department of Transportation—Pipelines and Hazardous Materials Safety Administration
- Department of Defense—US Army Corps of Engineers*

- Department of Defense—US Navy*
- Environmental Protection Agency
- National Aeronautics and Space Administration
- Department of Homeland Security—Federal Emergency Management Agency
- Department of Homeland Security—United States Coast Guard*

* Indicates active Arctic-focused R&D, related specifically to oil spills in ice-covered waters ICCOPR has evolved significantly over the past 20 years. In the early 1990s, after the passage of OPA90, in response to the *Exxon Valdez* oil spill, ICCOPR was actively engaged in fulfilling its responsibilities. However, by the late 1990s and into the early 2000s, activity waned, as funding for oil spill research diminished. Under current USCG leadership, in response to positive encouragement and increased scrutiny after the Deepwater Horizon incident, ICCOPR has been rejuvenated. The members are meeting quarterly, and have updated their charter and membership. They have developed a comprehensive website, and they are focused on addressing their mission and meeting their obligations as defined in OPA90. USARC recognizes and applauds this significant improvement in interagency effort, as led by the USCG.

The website includes a summary of related R&D: http://www. iccopr.uscg.gov/apex/f?p=118:366:1376080403462271.

The website recently added a "High Latitudes" link, to provide some of the resources that the ICCOPR uses to help identify and support research issues related to cold weather response: http://www.iccopr.uscg.gov/apex /f?p=118: 357:3718783452980200.

Other Arctic Marine Oil Spill R&D Efforts

Nonprofit Organizations

Oil Spill Recovery Institute (OSRI)

Authorized by the United States Congress through the Oil Pollution Act of 1990, the purpose of the Prince William Sound (PWS) Oil Spill Recovery Institute (OSRI) (http://www. pws-osri.org) is to support research, education, and demonstration projects designed to respond to and understand the effects of oil spills in the Arctic and sub-Arctic marine environments.

The OSRI 2011–2015 Research Plan (http://www. pws-osri.org/business/science_ plan.pdf) provides detailed program goals and objectives in four major areas:

a) *Understand.* Attain an interdisciplinary understanding of the fate and effects of spilled oil in Arctic and sub-Arctic marine environments, and the recovery of those environments following a spill.
b) *Respond.* Enhance the ability of oil spill responders to mitigate impacts of spills in Arctic and sub-Arctic marine environments.
c) *Inform.* Disseminate information and educate the public on the issues of oil spill prevention, response, and impacts.
d) *Partner* (as per efforts associated with the Interagency Arctic Research Policy Committee).

The OSRI website provides a comprehensive list of all the projects that have been supported, by year and category of research. Recent projects of particular interest in the Arctic include:

1) Support for NOAA's Arctic Environmental Response Management Application (ERMA®) (see NOAA section for more on ERMA®)
2) Detection and mapping of oil spills under sea ice (DAMOS), Scottish Association for Maine Science, and Woods Hole Oceanographic Institution (2011 and 2012). This project, which includes tests at CRREL's Geophysical Research Facility, demonstrates that oil can be found under ice with existing technologies. Off-the-shelf instruments, including cameras and sonar systems, were mounted on an AUV or ROV.
3) Oil in Ice: Transport, Fate, and Potential Exposure, University of Alaska, Fairbanks (2009, 2008). This project was a portion of a larger project funded by the CRRC to examine the transport of various components of hydrocarbons through sea ice, to examine the biodegradation potential of those components, and to model the combined results.
4) Support for the National Academy of Sciences study, "Responding to Oil Spills in the Arctic Marine Environment."

Industry (US and Non-US)

Alaska Clean Seas (ACS)

ACS provides response services to the Alaska North Slope Crude Oil Producers and the first 167 miles of the Trans-Alaska Pipeline System. The company is a not-for-profit oil removal response organization that began operations in 1979. Today, ACS serves 10 member companies. ACS trains responders in Alaska and at Ohmsett and CRREL facilities and at the Prince William Sound Community College in Valdez, Alaska. ACS participates in research to improve equipment needed to respond to spills in the Arctic. Recent research efforts involved using ground-penetrating radar to detect oil under the ice and in developing a helicopter-based oil-underice detection system. Currently, ACS is in the design phase of an oil and ice-testing tank at Prudhoe Bay, Alaska, in partnership with CRREL. The design criteria include nonferrous construction to minimize electromagnetic interference for testing oil detection antennae and sufficient depth for testing ROV/AOV based systems.

Joint Industry Programs (JIP)

For oil companies developing high-latitude natural resources, “stewardship” means investing in research to improve the oil spill response technologies and methods for use in the Arctic region. The oil companies’ first priority must be prevention of oil spills. However, in the event of a spill, it is essential to have proven techniques and methodologies to reduce environmental damage. If an oil company uses company funds to develop a technique, the result is often proprietary, though results are sometimes presented at related conferences. Industry often supports Joint Industry Programs (JIPs), and depending on the nature of the program, the results may be presented publicly. JIPs are funding mechanisms for the oil industry and for government agencies (such as DOI) to leverage their research dollars to find a solution to a common problem.

Oil spill prevention remains a priority for the oil and gas industry and the ability to prevent and respond to oil spills is essential to achieve licence to operate. Substantive technological advances to detect, contain, and clean up spills in Arctic environments have been made by industry in the past decade. Uniting efforts and knowledge through a JIP increases opportunities to test equipment, conduct field experiments, develop oil spill response technologies and methodologies, and raise awareness of existing industry oil spill response capabilities in the Arctic region. Sharing knowledge, not only within the industry, but also with authorities, academic institutions, and nongovernmental institutions is crucial.

American Petroleum Institute and the Joint Industry Program on Oil Spill Recovery in Ice: *Spill Response in the Arctic Offshore*

Published in February 2012, this report (http://www.api. org/~/media /Files/EHS/Clean_Water/Oil_Spill_Prevention/ Spill-Response-in-the-Arctic-Offshore.pdf) describes some of the challenges of responding to oil spills in the offshore Arctic, and discusses how industry has used results from research, technology development, and experimentation to respond to spills. The report was authored by consultants from SL Ross Environmental Research, DF Dickins Associates, and Polaris Applied Sciences, two of which are described below in the section "Private Consultants."

International Association of Oil & Gas Producers: Arctic Oil Spill Response Technology—Joint Industry Program

To build on existing research and improve the technologies and methodologies for Arctic spill response, the oil and gas industry established an Arctic Oil Spill Response Technology—Joint Industry Program (Arctic Oil Spill Response JIP) in January 2012, managed by the International Association of Oil & Gas Producers (OGP). The goal of this effort is to further improve oil spill response technologies through international research programs aimed at enhancing industry knowledge and capabilities in the areas of Arctic oil spill response. Nine oil companies are sponsoring the program: BP, Chevron, ConocoPhillips, Eni, ExxonMobil, North Caspian Oil Company, Shell, Statoil, and Total. All results from the Arctic Oil Spill Response Technology JIP will be published and made publicly available. Research initiatives under this JIP include:

- *Fate of dispersed oil under ic*e. Develop a numerical model capable of predicting the fate of a dispersed oil plume that develops under ice.
- *Dispersant testing under realistic conditions*. Define the operational limits of chemical dispersant and mineral fines in Arctic marine waters.
- *Environmental impacts of Arctic spills and Arctic spill response technologies*. Identify and conduct the research necessary to improve the knowledge base that supports NEBA (net environmental benefit analysis) and decision-making.

- *Trajectory modeling in ice.* Produce a verified/validated model that combines ice movement and spilled oil in ice to advance the state of knowledge of oil spill trajectory modeling in ice.
- *Oil spill detection and monitoring in low visibility and ice.* Identify and conduct the research necessary advance oil spill remote-sensing and mapping capabilities and technologies in darkness and low visibility, in broken ice, and under ice. Detect and track plumes that will develop if dispersants are used to control continuous subsea releases.
- *Mechanical recovery in ice.* Identify new approaches to facilitate the development of new or improved mechanical recovery equipment for use in the Arctic.
- *In situ burning state of knowledge.* Prepare materials to raise the awareness of industry, regulators and external stakeholders of the significant body of knowledge that currently exists on all aspects of in situ burning.
- *Aerial ignition systems.* Provide technology improvement that delivers a safe, reliable, and precise means of aerial ignition and improve oil slick targeting to support use of in situ burning.
- *Chemical herders to expand in situ burning window of opportunity.* Define the operational limits of chemical herders to allow in situ burning in open water and among broken ice.
- *Field research.* Evaluation of countermeasure technologies and conduct related research in a field setting.

The JIP projects, identified as "OSRT-JIP," are listed in an online table that accompanies this white paper (http://www.arctic.gov/publications /oil_spills_tableA.pdf). More information can be found at http://www.ogp.org. uk/news/press-releases/industry-programme-to-strengthen-arctic.

SINTEF: The Foundation for Scientific and Industrial Research (Norway)

SINTEF is a broadly based, multidisciplinary research organization that partners with the Norwegian University of Science and Technology (NTNU) in Trondheim, and collaborates with the University of Oslo. SINTEF offers services based on their experience with modeling, testing of oils in their laboratories and large-scale facilities, and numerous, real-time oil spill studies and analyses. The SINTEF SeaLab is specifically designed for oil spill R&D and includes laboratories and mesoscale facilities for conducting experiments associated with development of underwater production of oil, as well as problems related to oil spills for ice-infested and coastal waters with complex current patterns. For more information see http://www.sintef.no/home /Materials-and-Chemistry/Marine-Environmental-Technology/Projects-and-News/ SINTEF-SeaLab.

SINTEF was the lead on a Joint Industry Program on Oil in Ice with funding from Total, Statoil, ConocoPhillips, Shell, Chevron, and AGIP KCO. This JIP was established in 2006 and completed in 2009. It considered oil spill response techniques for Arctic waters and the fate and behavior of oil spills in ice and under cold-water conditions. Key findings of the Oil in Ice JIP include:

- The research program provided a valuable knowledge base to plan, implement, and further improve oil spill response in ice-covered waters.
- Each response tool evaluated during the program demonstrated some merit in responding to an oil spill an Arctic environment.
- The availability of all the response options was considered as being the key to a successful oil spill response operation in Arctic conditions.
- A systematic way to predict the operational time frame for various response options was identified, with implications for the efficiency of spill response.
- Large-scale field experiments supported results from a number of small- and medium-scale laboratory experiments.
- Laboratory and field experiments suggested that in situ burning and chemical dispersion may be appropriate response methods.
- The presence of cold water and ice can enhance response effectiveness by limiting the spread of oil and slowing the weathering process.
- The window of opportunity for in situ burning and dispersant operations in ice-covered waters can significantly increase compared to an open-water scenario under certain circumstances.
- New technologies for mechanical oil spill recovery and dispersant application that, when combined with a large set of test data, will improve response planning and response operations.

The full report on the Oil in Ice JIP can be found at http://www. sintef.no/project/JIP_Oil_In_Ice/Dokumenter/publications/ JIP-rep-no-32-Summary-report.pdf.

Barrow/JIp (Alaska)

From 2009–2011, four companies (Shell Exploration and Production Company, ExxonMobil Upstream Research Company, Statoil Petroleum ASA, and ConocoPhillips) teamed to study the "Toxicology and Biodegradation of Crude and Dispersed Oil in the Arctic Marine Environment." The effort was managed by Jack Word of NewFields and Robert Perkins of the UAF's Institute of Northern Engineering. A Powerpoint presentation, summarizing the results, can be found online at http://www.newfields.com/dl /alaskaworkgroup/JIP Program 2009-2011/JIP Summary Updates and Presentations/UAF-NewFields-Toxicology and Biodegradation in Arctic1feb 2011.pptx.

Statoil (Norway)

Statoil completed a large research program in 2010 looking at the technologies for oil spill response in cold and icy conditions. Existing tools (e.g., chemical herders, fire resistant booms, remote-sensing systems) as well as newly developed technologies (e.g., mechanical oil recovery systems, dispersant application systems) were tested in different environmental scenarios. According to Statoil, the key findings from the program are: (1) all the response techniques could be used in responding to an oil spill in an Arctic environment, depending upon conditions; (2) the time window for use of in situ burning and the use of dispersants in ice-covered waters can increase significantly compared with an open water scenario because of the presence of cold water and ice (ice limits the spread of oil, slowing down the weathering process); and (3) having all response options available is considered to be the key to a successful oil spill response operation under Arctic conditions.

Canada

Centre for Offshore Oil, Gas, and Energy Research (Cooger) Canadian Federal Agency)

The Department of Fisheries and Oceans Canada (DFO) established the Centre for Offshore Oil, Gas, and Energy Research (COOGER) in 2002 to coordinate its nationwide research into the environmental and oceanographic impacts of offshore petroleum exploration, production, and transportation (see http://www.dfo-mpo.gc.ca/science/coe-cde/cooger- crpgee/index-eng.htm). COOGER's mandate was expanded in 2009 to include ocean renewable energy (i.e., energy conversion from tide, wind, and waves) in response to emerging technological developments, so that Canada can meet its future energy needs in an environmentally responsible manner. By building on existing regional expertise and infrastructure to coordinate research at a national scale, COOGER addresses its departmental mandate and industry needs by providing scientific knowledge for use to ensure safe and environment ally sound management of offshore oil, gas, and renewable energy operations.

Based at the Bedford Institute of Oceanography in Dartmouth, Nova Scotia, COOGER maintains an extensive oil spill research program in partnership with national and international agencies, including the Canadian Coast Guard, US EPA, NOAA, BSEE, the Coastal Response Research Center (University of New Hampshire), Cedre (France), SINTEF, Akvaplan-niva (Norway), and others in the private sector through Joint Industry Program (JIP) agreements under auspices such as the International Association of Oil & Gas Producers. With the expansion of frontier oil and gas development offshore of Canada, a primary focus of COOGER's research is to understand the fate, behavior, and effects of oil spills, and to assess the efficacy of countermeasure technologies.

COOGER operates advanced chemistry and biology laboratories to conduct bench-scale research as well as to provide analytical support to oil spill response operations. The group maintains state-of-the-art instrumentation for petroleum hydrocarbon analysis that include Iatroscans, gas chromatography-flame ionization detectors (GC-FID), and GC-mass spectrometers (GC-MS), which enable fingerprinting for the identification of source oils and the tracking of physicochemical and biological processes on oil spilled at sea and/or stranded within shoreline sediments.

COOGER has led experimental field trials involving the controlled release of oil within Canada, in addition to collaborating internationally with others in the United States, Norway, France, and the Netherlands. These studies have established operational guidelines pertaining to the application of bioremediation, phytoremediation, and natural attenuation strategies for the remediation of oil spill impacted sites. COOGER has also been involved in the development of standard techniques for the identification of operational endpoints based on habitat recovery for oil spill response operations.

A key aspect of COOGER's research is centered on its wave tank facility that was designed specifically for the evaluation of chemical dispersant effectiveness (under different wave energy regimes) and the potential biological effects associated with their application. Recent studies have also been focused on the interaction of oil spilled at sea with suspended particulate material (SPM) and the processes controlling oil-mineral aggregate (OMA) formation that promote both dispersion and biodegradation of oil spilled at sea. The aim of this research is to provide information for the support of decision making on the spill response option to be taken based on net environmental benefit analysis (NEBA).

COOGER has conducted a number of research studies pertaining to oil spills in the Arctic. They have included shoreline studies on the bioremediation (by nutrient addition and tilling) and surf-washing to remove residual oil stranded within supra- and intertidal sediments. Recent field studies have included a program in the Gulf of St. Lawrence in the presence of broken ice during the winter to evaluate the feasibility of enhanced OMA formation as a spill response strategy for use under Arctic conditions. Arctic field experiments are also being conducted to determine the capacity of natural bacteria to degrade oil and chemically dispersed oil.

COOGER's expertise was called upon by the US government in 2010 to assist with monitoring the use of chemical dispersant during the Deepwater Horizon spill. The Centre's Executive Director, Dr. Kenneth Lee, is an international expert in oil spill research and sits on a number distinguished committees including the US National Academy of Sciences/ National Research Council Committees on the "Effects of the Deepwater Horizon Mississippi Canyon-252 Oil Spill on Ecosystem Services in the Gulf of Mexico" and "Responding to Oil Spills in Arctic Environments."

Environment Canada (EC)

Environment Canada's (EC) mission is to protect the environment, conserve the country's natural heritage, and to provide weather and meteorological information to Canadians. The diverse organization collaborates with research partners on a variety of initiatives and participates in the Arctic Council. Agencies within EC that have conducted oil spill research include: Conservation and Protection Agency, Emergencies Engineering Division, Emergencies Science Division, and Western Office, Technology Development Branch.

In March 1978, EC began the Arctic and Marine Oil Spill Program (AMOP) to improve the knowledge base and technology for responding to Arctic and marine oil spills. The AMOP Technical Seminar on Environmental Contamination and Response is organized annually by the Emergencies Science and Technology Section (ESTS) and covers research and development on a variety of topics related to environmental emergencies caused by spilled hazardous materials. For more than 30 years, the Section has run a continuing national program of R&D on:

- Properties, behavior, detection, measurement, and effects of spilled hazardous materials
- Modeling and remote sensing of spilled hazardous materials
- Spill countermeasures: evaluation, effectiveness, effects, and environmental benefits of mechanical and chemical treating agents
- Shoreline impact and restoration, specifically development of the Shoreline Cleanup Assessment Technique (SCAT)

The 34th AMOP Technical Seminar was held October 4–6, 2011 in Banff, Alberta, with more than 100 papers and posters presented. The 35th technical seminar was held on June 5–7, 2012, in Vancouver, British Columbia. The program is available at http://www.ec.gc.ca/scitech/default. asp?lang=En&n= A0477462-1.

A unique feature of EC-supported R&D is that results are related to actual spill incidents, providing assistance to spill responders and conversely feedback to the researchers on the application of their work. R&D priorities are set and assessed by committees of representatives from all levels of government as well as international government agencies. Technology transfer is an important component of the program and the group provides operational guides, manuals, and training as well as some aspects of contingency planning.

Over the past decade, ECs budget for oil spill research has been lower, and the level of activity has been reduced. EC continues to maintain sophisticated laboratories for contaminant and sample analysis, but budget reductions have eliminated many other field programs and research activities.

University of Manitoba Sea-Ice Environmental Research Facility (Serf) (Canada)

The Sea-Ice Environmental Research Facility (SERF) has a covered tank that relies on ambient temperature for sea ice growth. The existing tank is 60 feet long, 30 feet wide, and 8 feet deep, and is primarily used to grow sea ice under various controlled conditions, conduct mesocosm-scale studies to enhance the fundamental understanding of how sea ice forms and melts on polar oceans, and gain insight into the processes that regulate the exchange of energy and matter between the ocean and atmosphere. Detailed studies in collaboration with field measurements are being conducted to improve ability to predict the impact of the rapid sea ice loss on the marine ecosystem, on

Arctic and global climates, on transport and biogeochemical cycles of greenhouse gases and contaminants, and on the human use of sea ice. A second tank for conducting oil and ice studies is planned. See http://home.cc.umanitoba.ca/~wangf/serf.

Private Consultants

Private consultants are providing support to oil companies and manufactures of oil spill response equipment. The results of their work are not typically in the public domain, unless the work is associated with a Joint Industry Program. As an example, two firms active in JIP efforts are DF Dickins Associates and SL Ross Environmental Research.

DF Dickins Associates LTD (USA)

Dickins specialties include engineering and environmental studies associated with offshore oil exploration and development, coastal mine sites, and marine transportation. More recently work has been in the area of remote sensing of oil spills under ice. See http://www.dfdickins.com.

Sl Ross Environmental Research (Canada)

SL Ross investigates a variety of oil spill countermeasures in the laboratory and field, including dispersant testing, in situ burning evaluations, testing of skimmers and booms, development of equipment for shoreline cleanup and disposal, sorbent testing and evaluation, and behavior of oil spills. SL Ross has a laboratory facility, which contains a 10 m long wind/ wave tank. SL Ross also uses Ohmsett, the refrigerated facilities at CRREL, and a custom-built wave tank located on the North Slope of Alaska. SL Ross staff have also conducted several major field studies in the open ocean and in ice involving evaluations of dispersants, skimmers, booms, burning, and oil behavior. See http://www.slross.com.

Other Current Activities of Interest and Published R&D Plans

Arctic Council

On an international level, the Arctic Council Ministers, at their May 2011 meeting, created a task force to examine the possibility of bringing the Arctic States together to address oil spill preparedness and response. The United

States initiated this proposal, and agreed to co-chair the Task Force with Norway and Russia. The impetus behind the US proposal was twofold:

- That the Arctic Council take the lead in ensuring that the Arctic States are well prepared for, and able to smoothly coordinate collective response to, possible oil spills in connection with increasing human activity in the region (e.g., offshore development, shipping, pipeline construction)
- That the lessons of the Deepwater Horizon event inform the Arctic States in order to improve international coordination for response

The Task Force decided to negotiate an instrument—legally binding—on oil spill preparedness and response based on the 1990 International Convention on Oil Pollution Preparedness Response and Cooperation (OPRC), administered by the International Maritime Organization (IMO), to which all eight Arctic States would be party. The instrument, which would be a regional multilateral agreement under the OPRC (Article 10), is likely to establish a cooperative framework similar to the Arctic Search and Rescue Agreement. The Task Force is likely to recommend to the Ministers whether and how best to pursue future work in this area, such as possible substantive annexes to the framework instrument that address technical aspects oil spill preparedness and response.

Norway hosted the first negotiating session in Oslo in October 2011, Russia hosted the second session in St. Petersburg in December 2011, and the United States hosted the third session in Anchorage in March 2012. A fourth session was hosted by Finland in June 2012.

At the time of publication of this white paper, the most recent development was the fifth meeting, in Reykjavik, Iceland, during the week of October 8, 2012, which involved delegates from the eight Arctic Council member states. The Task Force concluded negotiations on the instrument, with only minor technical issues left to resolve. It is expected that the agreement will be signed by the Ministers in May 2013.

At their May 2011 meeting, the Ministers also directed the Arctic Council working groups, led by the Emergency Preparedness, Prevention, and Response (EPPR) working group (http://eppr.arctic-council.org), to develop recommendations and best practices for oil spill prevention. This work is occurring in close consultation with the Task Force, and will be submitted, together with the instrument, to the May 2013 Ministerial meeting.

US National Research Council

National Research Council (Ocean Studies Board, Polar Research Board, Marine Board) is currently coordinating a study titled "Responding to Oil Spills in Arctic Environments." According the Statement of Task, the NRC will assess the current state of the science regarding oil spill response and environmental assessment in the Arctic region (with a specific focus on the regions north of the Bering Strait), with emphasis on potential impacts in US waters. The study is currently supported by several entities, including USCG, NOAA, BOEM, BSEE, US Arctic Research Commission, American Petroleum Institute, the State of Alaska, and others. The committee has been formed, and the initial meeting of this group is scheduled for December 17, 2012, in Washington, DC. The target date to complete the report is 2014 (24 months after the initial committee meeting).

US Congress

In January 2011, Senator Begich re-introduced two bills related to Arctic oil spills and research:

1) S.203 | A bill to direct the Administrator of NOAA to institute research into the special circumstances associated with oil spill prevention and response in Arctic waters, including assessment of impacts on Arctic marine mammals and other wildlife, marine debris research and removal, and risk assessment, and for other purposes. SPONSOR: Sen. Begich, Mark [AK] (introduced 1/26/2011). CO-SPONSORS: None. COMMITTEES: Senate Commerce, Science, and Transportation. LATEST MAJOR ACTION: 1/26/2011 Referred to Senate committee. STATUS: Read twice and referred to the Committee on Commerce, Science, and Transportation.
2) S.204 | A bill to amend the Oil Pollution Act of 1990 to permit funds in the Oil Spill Liability Trust to be used by NOAA, the USCG, and other federal agencies for certain research, prevention, and response capabilities with respect to discharges of oil, for environmental studies, and for grant programs to communities affected by oil spills on the Outer Continental Shelf, and to provide funding for such uses and for other purposes. SPONSOR: Sen Begich, Mark [AK] (introduced 1/26/2011). CO-SPONSORS: None. COMMITTEES: Senate Finance. LATEST MAJOR ACTION: 1/26/2011 Referred to Senate committee. STATUS: Read twice and referred to the Committee on Finance.

State of Alaska: Department of Environmental Conservation's Division of Spill Prevention And Response

The State of Alaska has substantial experience with oil and gas development in the Arctic. Alaska has one of the largest Arctic oil fields and has conducted substantial offshore Arctic drilling, approximately 70 wells to date, with a great deal of experience in the use of offshore platforms in ice conditions. Four offshore production units are in operation in Alaska's Arctic. There are also 16 offshore platforms in ice-infested Cook Inlet waters, the first of which became operational in 1964. This Arctic and sub-Arctic experience has led to a sophisticated regulatory framework for oil spill prevention and response in ice-covered waters, and a robust preparedness and response program. The State of Alaska is equipped, and has a drill and exercise program. This state and industry experience offers a useful perspective from which to shape future recommendations for additional research in ice covered waters.

Based on GAO report 11-319, published in March 2011 (available at http://www.gao.gov/assets/320/317056.pdf), the State of Alaska has had a spill prevention and response program since the 1970s. As a result of judgments entered in the criminal cases for the *Exxon Valdez* oil spill, funds were provided to the State of Alaska to enhance the ability of the state and industry to respond to oil spills. Since 1989, a total of $2.5M has been provided to the Alaska Department of Environmental Conservation (DEC) for projects under this program. The funds have been used for research programs directed toward the prevention, containment, cleanup, and amelioration of oil spills in Alaska.

In cooperation with other stakeholders, DEC has supported over 30 research and development projects dealing with subjects such as cleanup technology, nonmechanical response techniques, the fate and effects of spilled oil, oil spill contingency planning and preparedness, spill response training, incident management systems, and spill prevention. Research supported by DEC has been conducted by Alaskan-based oil-spill response cooperatives, private consultants, universities, and other state and federal agencies. Details of the research can be found at http://dec.alaska.gov/spar/perp/r_d/research_list.htm.

University of Alaska Fairbanks (UAF)

The University of Alaska Fairbanks (UAF) submitted a pre-proposal to NSF to establish a Science and Technology Center (STC) for Oil Spill Prevention and Preparedness in the Arctic. UAF was invited to provide a full

Center proposal by Feb 3, 2012. The proposal was submitted, but it was not subsequently selected for support.

Regardless, the concept is worth describing. The STC is envisioned as a center-based approach, partnering UAF with CRRC and others, to perform the necessary research and development to prepare for potential spills in the Arctic, including understanding impacts to the ecosystem and developing techniques to mitigate spills should they occur. New techniques for detecting, tracking, and monitoring the movement and fate of oil will be required. In addition, basic research into the current ecosystem baseline and expected behavior of oil in the presence of ice will be required to advance knowledge of the arctic environment. The Center aims to: (1) identify knowledge gaps through stakeholders and subject experts, (2) sponsor fundamental research in science relevant to the Arctic environment, (3) sponsor applied research with joint funding from industry as well as state and federal agencies, (4) administer fellowships and grants related to education in Arctic science and technology, (5) transfer knowledge gained via workshops, meetings, literature, training, and electronic media, and (6) create feedback mechanisms to continuously identify knowledge gaps and evaluate efficacy of results.

Recent Workshop: Oil Spill in Sea Ice— Past, Present, and Future

A workshop was held by Istituto Geografico Polare "Silvio Zavatti" on September 20–23, 2011, in Fermo, Italy (http:// www.oilspillsinseaice.net). Thirty-tree delegates representing 12 countries addressed the question: "How can we design an effective, integrated system for dealing with every aspect of a potential accident in ice-covered waters which involves the release of oil?" A key outcome of the workshop was a "Fermo Statement," which has been submitted to the Arctic Council for consideration by the by the Arctic Task Force set up by the Council to examine protection of the Arctic against oil pollution (see above under "Arctic Council"). In 2013, the papers given at the workshop will be published in a proceedings volume and, after refereeing, in a special issue of the journal *Cold Regions Science and Technology*. The delegates at the workshop summarized their points into 12 themes:

1) How best to stop a blowout
2) How to model oil spread
3) Tracking oil spills
4) Problems with in situ burning
5) The role of dispersants
6) The physics of large-scale oil entrapment

7) The biological consequences of oil spills
8) The rapidity of environmental change
9) Data sharing and management
10) A rapid scientific response
11) Delivery of the oil to the ice underside
12) The natural background

USARC RECOMMENDATIONS

The United States Arctic Research Commission (USARC) has long supported the existence of basic and applied research programs to improve methods to prevent and respond to oil spills in the Arctic region, and to understand the fate and effects of oil released into the environment.

In 2004, USARC published *Advancing Oil Spill Response in Ice-Covered Waters* (http://www.arctic.gov/publications/ oil_in_ice.html), which defined a program and identified research and development projects to improve the ability of responders to address accidental oil spills in fresh- or salt water marine environments where ice is present. This response program included spills that occur on or beneath solid, stable ice extending out from shore, as well as spills in areas of drifting ice floes, and on ice-covered shorelines.

In 2010, USARC published a white paper titled *USARC Recommends Steps to Expanded US Funding for Arctic/ Subarctic Oil Spill Research* (http://www.arctic.gov/ publications/oil_spill_wp.html) that recommended ways to invigorate oil spill research in the United States. Several of those recommendations have been implemented, or are in the process of being implemented. For example, the Interagency Coordinating Committee on Ocean Pollution Research (ICCOPR) is currently updating its research and technology plan for the first time since 1997, and NOAA is now co-chairing ICCOPR. Second, government and industry funding has enabled greater research on the basic ecological structure of the marine environment in the Chukchi and Beaufort Seas. Funds have been provided by BOEM, a consortium of oil and gas companies, and the North Pacific Research Board.

This white paper updates the 2010 publication and offers the following recommendations to encourage those who are undertaking research to consider these areas for additional work. We do not prioritize these recommendations, and we also realize that, in some cases, these recommendations are presently being acted upon. We look forward to the implementation of recommendations, peer-reviewed publications that stem from work performed in these areas, and, most importantly, to the practical application of knowledge gained in real-world situations.

Spill Delineation and Mitigation, Including Containment and Countermeasures

- Develop spilled oil detection and mapping techniques that are independent of ice conditions, are rapidly deployable, and can be quickly interpreted in the field. Specific techniques that need enhancement are satellite and airborne sensors for cursory assessment, with refined delineation via airborne and underwater systems.
- Field test and train various techniques by the responders to assure familiarity with the process before it has to be deployed for a spill.
- Place greater emphasis on R&D for source control, and optimize Arctic vessel lightering, Arctic tank and pipeline evacuation methods, and Arctic well capping/containment.
- Develop a database that collects, analyzes, and then summarizes data from actual Arctic oil spills. These data would be used to ground truth the range of operating conditions and equipment/human factor limitations that have an actual effect on oil spill prevention planning and oil spill response improvements. To this end, BOEM has initiated a study entitled "Oil Spill Occurrence Estimators for Onshore Alaska North Slope Crude and Refined Oil Spills" and another called "Loss of Well Control Occurrence and Size Estimators for the Alaska OCS."

Oil Spill Response Technologies for Cleanup and Recovery of Oil

- Improve performance of mechanical oil recovery systems deployed in ice-infested water (especially in spring broken ice and fall freeze-up conditions).
- Test in situ burning, which can be a viable option for minimizing the volume of oil. Chemical herders have been effective in scavenging the oil in broken ice, expanding the window of opportunity for recovery, and improving the cleanup process. Additional research is needed to improve the herder's performance in Arctic conditions including waves and cold water, and analyze their toxicity.

- Develop robots and ROV/AOV to assist in the cleanup process, which can be very useful during times when human response is not possible.
- Test methods to recover oil trapped under ice. Improve tools to measure and map oil spill thickness to identify areas of pooled oil that may be thick enough to collect or burn in situ.
- Evaluate dispersant efficacy and impacts in Arctic conditions. Dispersants were used extensively in the Deepwater Horizon cleanup, with considerable research done and still underway to assess their effectiveness and toxicity. The impacts of Arctic environmental conditions on the effectiveness of the dispersants should be assessed, including their ability to disperse oil by injecting at the source of submerged leaks and their potential impacts on the local wildlife and its consumers.

Data Management Tools Currently Being Developed, and the Fate of Oil and Its Effects on the Environment

- Quantify the environmental effects of in situ burning, including the smoke and residue, so they can be included in the response assessment analysis, to assist in deciding the best approach for remediation as a function of in situ conditions.
- Define the effects of the dispersed oil on the Arctic ecosystems, with assessments of the toxicity of dispersed oil and dispersants on benthic flora and fauna, in particular, marine mammals and sea birds.
- Assess environmentally sensitive areas. Much of the Chukchi and Beaufort coasts have been identified as areas of heightened ecological significance. Much of the shoreline zones are shallow and composed of ice-rich permafrost soils. The impacts of both crude oil and dispersed oil deposition on these shorelines and the cleanup techniques available should be assessed. BOEM has some ongoing studies of such areas, but more are needed.
- Conduct additional research on the efficacy, and the human health, biological, and toxicological effects of chemical herders.

General

- Provide federal agencies with basic data, as well as summary reports, from industry research projects. USARC recommends data sharing as a cost-effective mechanism to increase federal response capacity.
- Analyze the "Response Gap" in areas that will soon be developed, to statistically quantify current response limitations. A "Response Gap" is the period of time when oil spill response is not possible because one or more limiting factors preventing an effective response (e.g., weather, sea ice, wave height, darkness, and extreme cold). Response Gap analyses have been performed in several areas experiencing or anticipating offshore oil and gas development, such as Davis Strait and Cook Inlet. They provide critical insight into needed planning efforts and identify the weakest link in the response chain that needs to be addressed.
- Test existing technologies in the specific ice conditions of concern (i.e., spring breakup, deteriorating spring ice, and fall freeze-up). Actual oil spill experiments and field trials are needed in the Beaufort and Chukchi Seas, employing local oil samples to improve tactics, improve response equipment and train personnel.
- Create a clear, concise description of human safety limitations (for response operations) to be included in oil spill response plans.
- Assess prior R&D projects to determine if they produced useful results. Were measurable advances in oil spill response techniques achieved, and in actual field response capabilities?
- Reveal, in agency budgets, the funding allocated for research on oil spills in ice-covered waters. This funding should increase, given the emphasis on the frontier prospects in the Arctic.
- Establish regulatory benchmarks and Best Available Technology (BAT) regulatory requirements for oil spills. These requirements would parallel those set for air pollution controls to incentivize greater private funding of R&D, which is currently lacking.
- Use research results to help create a BAT database for Arctic response equipment. If such a database existed, and was required for the Alaska Department of Environmental Conservation (ADEC) and BSEE, then permit applicants and regulators would install the BAT, or explain why it is not technically or economically feasible.
- Create an Arctic Subcommittee under ICCOPR that would focus on Arctic spill prevention, preparedness, and response, and base it in

Alaska, in order to enable local stakeholders, oil spill cooperatives, industry, and the State of Alaska to consolidate knowledge and to seek consensus on Arctic research priorities.

- Fund NOAA for oil spill research and development from the Oil Spill Liability Trust Fund (OSLTF), in light of the fact that NOAA is the only federal agency with oil spill preparedness, response, and restoration responsibilities under OPA90 that does not receive a direct appropriation from the OSLTF.
- Increase either the amount of time that federal agencies are given to review permit requests, or the number of technical staff, given the breadth of oil spill R&D.
- Increase the use of peer review and greater transparency in publications. R&D results need to be reported accurately, including full disclosure on the scope of work, results, and what was concluded.
- Increase efforts to better communicate the state of knowledge, and the value, of oil spill research to the public and to the media.

In: Oil Spills in Arctic Waters
Editor: Nora Linwood
ISBN: 978-1-63321-580-1

Chapter 2

TABLE A: AN INVENTORY OF RESEARCH PROJECTS ON OIL SPILLS IN ARCTIC WATERS*

United States Arctic Research Commission and US Army Corps of Engineers Cold Regions Research and Engineering Laboratory

This matrix is a selection of Arctic oil spill research drawn largely from US government sources and supplemented by representative projects from other sources including foreign governments and privately funded programs. The collection of references are limited to those available in the public domain.

* This is an edited, reformatted and augmented version of an online table that accompanies the white paper found in Chapter 1, issued November 2012.

Research Initiatives	Description	Facility	PI	Funding	Year	Objective
Assessments and Workshops						
	Interagency Protocols for Immediate On-Scene Arctic Oils Spill Impact Science		NA	BOEM-AK	2012	Beaufort, Chukchi
	Arctic and Marine Oilspill Program (AMOP) Technical Seminar on Environmental Contamination and Response		NA	EC	2012	http://www.etc-cte.ec.gc.ca/news/conferences_e.html
	Spill Response in the Arctic Offshore		SL Ross, DF Dickins Associates, Polaris Applied Sciences	API- OSRI	2011	Prepared for the American Petroleum Institute and the Joint Industry Programme on Oil Spill Recovery in Ice http://www.api.org/~/media/Files/EHS/Clean_Water/Oil_Spill_Prevention/Spill-Response-in-the-Arctic-Offshore.ashx

Research Initiatives	Description	Facility	PI	Funding	Year	Objective
	Offshore Technology Conference- Arctic				*2011*	
	Joint Industry Program on Oil Spill Contingency for Arctic and Ice- Covered Waters: Summary Report		SINTEF, SL Ross, DF Dickins Associates	JIP SINTIF Oil In Ice	2010	Summary of the Joint Industry Program on Oil Spill Contingency for Arctic and Ice-Covered Waters (JIP on oil in ice. http://www.sintef.no/project/JIP_Oil_In_Ice/Dokumenter/publica tions/JIP-rep-no-32-Summary-report.pdf
	Beaufort Sea Oil Spills State of Knowledge Review and Identification of Key Issues		SL Ross, DF Dickins Associates, Envision Planning Solutions	Environmental Studies Research Funds	2010	Comprehensive review of the Arctic oil spill knowledge, including behavior and modeling, surveillance and monitoring, containment, recovery, in situ burning, dispersants, oil waste management, shoreline spill response and case histories. http://www.esrfunds.org/pdf/177.pdf

(Continued)

Research Initiatives	Description	Facility	PI	Funding	Year	Objective
	Advancing Oil Spill Response in Ice- Covered Waters				2009	http://www.arctic.gov/publications/ oil in ice.pdf
	Arctic and Marine Oilspill Program (AMOP) Technical Seminar on Environmental Contamination and Response				2009	Conference is every other year. http://www.ec.gc.ca/scitech/default. asp?lang=En&n=66A57AF7- 1
	International Oil in Ice Workshop 2007		SL Ross, DF Dickins Associates	MMS, ACS, Alaska Department of Environmental Conservation, Cook Inlet Spill Prevention and Response, OSRI, USCG	2008	Project is to organize and conduct an international workshop on recent advances in cleanup of oil spills in ice and cold climates.

Research Initiatives	Description	Facility	PI	Funding	Year	Objective
	Arctic Oil Spill Response Research and Development Program: A Decade of Achievement		MMS	MMS	2008	Report provides a comprehensive summary of activities and accomplishments of the (MMS) Arctic Oil Spill Response Research (OSRR) program. Significant accomplishments of the OSRR program include: • Detection of oil in and under ice • Oil spill thickness sensor • Mechanical containment and recovery in ice environments • In situ burn research • Dispersants in cold water/broken ice environments • Chemical herders • Operation of Ohmsett http://www.iccopr.uscg.gov/iccopr/i/files/MMSArcticResearch_2009.pdf

(Continued)

Research Initiatives	Description	Facility	PI	Funding	Year	Objective
	Short State-of-the-Art Report on Oil Spills in Ice-Infested Waters		SINTEF	JIP SINTIF Oil In Ice	2006	The objective of this report is to give a short and "to-the-point" presentation of the status, knowledge gaps, and research needs regarding oil spill response in Arctic and primarily in ice-infested waters. The report identifies technology gaps and recommendations are to be used to develop a scope of work for an R&D program to support petroleum exploration and production in Arctic areas. http://www.sintef.no /project/JIP_Oil_In_Ice/Dokumenter/publica tions/JIP-rep-no-1-State-of-the-art-2006-oil-in-ice.pdf
	Partnering in a Workshop to Determine the Scope of an Experimental Oil Spill in Pack Ice in Canada		SL Ross, DF Dickins Associates	MMS	2006	Conduct a planning workshop for an experimental oil spill in pack ice offshore Canada.

Research Initiatives	Description	Facility	PI	Funding	Year	Objective
	Transfer of Decanting Technology Research to Oil Spill Response Organizations and Regulators		SL Ross	MMS	2005	Distill research results collected during the past six years of decanting experiments (including the use of chemical demulsifiers to enhance water separation) into proposed guidelines for operational use and to transfer technology to responders and regulators.
	Workshop on the Issue of Prevention: What are the Next Challenges?		American Petroleum Institute	MMS	2005	International Oil Spill Conference (IOSC) is the pre-eminent gathering of oil spill experts from around the world.
	Project Canceled Summary Report of Activities at the Bureau of Ocean Energy Management, Regulation, and Enforcement Ohmsett Facility		ATRP Corporation	MMS	2005	The objective of this project was to prepare a series of technical summaries detailing the oil spill related research, testing, and training activities conducted at Ohmsett.

(Continued)

Research Initiatives	Description	Facility	PI	Funding	Year	Objective
	Global Challenges to Preparedness and Response Regimes Workshop		American Petroleum Institute	MMS	2003	International Oil Spill Conference (IOSC) is the pre-eminent gathering of oil spill experts from around the world.
	International Oil and Ice Workshop		DF Dickins Associates	MMS & ACS	2002	The workshop assembled experts on oil fate and behavior, Arctic oil spill response, ice environments, and Arctic oilfield development to present the leading edge technologies in a seminar and field setting.
	Alaska Arctic Workshop		NIST	MMS	1990	The objectives of the workshop were to: 1. Provide a public forum to describe existing research programs 2. Identify future research needs and priorities

Research Initiatives	Description	Facility	PI	Funding	Year	Objective
						3. Improve and to advance Arctic oil spill response capabilities 4. Present discussions of the state-of-the-art for all aspects of oil spill response under Arctic conditions Alaska Arctic Offshore Oil Spill Response Technology, November 29– December 1, 1988, Workshop Proceedings, National Institute of Standards and Technology Special Publication 762, April 1989.
Fate and Behavior of Oil						
	Physical Behavior of Oil in the Ocean		EC-ESD	MMS & EC	2002	Behavior and oil properties, particularly those of heavy oils. The study includes: buoyancy behavior, solubility, evaporation, dispersion, photo-oxidation, and emulsification
	Emulsions Formed at Sea and in Test Tanks		EC	MMS	2002	Evaluate the tendency of oils to emulsify at sea into unstable, mesostable, or stable emulsions

(Continued)

Research Initiatives	Description	Facility	PI	Funding	Year	Objective
	Testing at Ohmsett to Determine Optimum Times to Decant to Temporary Storage Devices		SL Ross	MMS	2001	Conduct full-scale tests of recovered oil/emulsion/water separation rates
	Fate and Behavior of Deepwater Subsea Oil Well Blowouts in the Gulf of Mexico		SL Ross	MMS	1997	Analyze the behavior of oil spilled during a deepwater blowout
	Development of a Portable Oil Analysis Kit for Responders		EC-ESD	MMS & EC	1992	Develop a field kit to measure oil properties

Research Initiatives	Description	Facility	PI	Funding	Year	Objective
Environmental Effects						
Proposed	Environmental impacts of Arctic spills and their response			OSRT-JIP	2012	Provide a robust information base that will support the use of net environmental benefit analysis (NEBA) for Arctic oil spill environmental impact assessments and response decision-making
Ongoing	Environmental Response Management Application (ERMA)		UNH	NOAA, UNH, EPA,USCG and Department of Interior,	2012	Online mapping tool that integrates both static and real-time data, such as Environmental Sensitivity Index (ESI) maps, ship locations, weather, and ocean currents, in a centralized, easyto-use format for environmental responders and decision makers. http://response. restoration.noaa.gov/erma_portal
	Aggregate Effects Research & Environmental Mitigation Monitoring of Oil Operations in the Vicinity of Nuiqsut			BOEM-AK	2012	Beaufort

(Continued)

Research Initiatives	Description	Facility	PI	Funding	Year	Objective
Detection, Mapping and Tracking						
Proposed	Trajectory modeling in ice			OSRT-JIP	2012	Create or adapt an existing numerical model that is capable of modeling the trajectory of spilled oil in various ice concentration
	Oil spill detection and monitoring in low visibility and ice			OSRT-JIP	2012	To advance and expand the oil and gas industry's oil spill remote sensing and mapping capabilities and technologies in Arctic conditions
Reports on Remote Sensing	Project P5: Remote Sensing Summary Report		DF Dickins Associates	JIP SINTIF Oil In Ice	2010	The main objective s of this report are threefold: (1) summarize the findings of the primary JIP remote sensing activities from 2007 to 2009,(2) draw conclusions recommending the most effective combination sensors and systems based on information gathered through the JIP,

Research Initiatives	Description	Facility	PI	Funding	Year	Objective
						including the initial technology screening and assessment and field experiments offshore and on Svalbard, and (3) highlight several evolving technologies and/or research that could enhance industry's capabilities in the next few years. http://www.sintef.no/project/JIP_Oil_In_Ice/Dokumenter/publications/JIP-rep-no-30-RSSummaryFinal.pdf
	Remote Sensing Technology Review and Screening		DF DickinsAssoci ates & Norconsult	SINTEF Oil in Ice JIP - AGIP KCO, Chevron,Conoco Phillips,Shell, Statoil & Total	2009	Spill detection and mapping are particularly important for Arctic spills as oil may be hidden from view under snow and ice during periods of almost total darkness. Very little is known about the capabilities of existing remote sensing systems—airborne, surface, and satellite—when faced with an accidental spill in a specific ice condition ranging from very open

(Continued)

Research Initiatives	Description	Facility	PI	Funding	Year	Objective
						drift ice (1–3/10) to very close winter pack ice (9/10). This report considers the likely performance of different sensors in a range of ice types based on their signal attributes and identifies the most likely systems for future field testing in the JIP such as airborne multispectral systems, synthetic aperture radar, ground penetrating radar, and trained dogs. http://www.sintef.no/project/JIP_Oil_In_Ice/Dokumenter/publications/JIP-rep-no-22-RSScreening-Final.pdf
Underwater	Underwater sensors to detect and map oil spills under ice	CRREL	OSRI	WHOI & SAMS	2012	Explored the possibility of using upward-looking capabilities to detect and map oil spills under ice. The suite of sensors, to include cameras, sonar and lasers, were attached to a submersible trolley, which was pulled under ice.

Research Initiatives	Description	Facility	PI	Funding	Year	Objective
Surface/ Instrumentation	Measurement of Methane Emission from Oil Spill Experiments at Svea Test Site, Svalbard, April 2007		Shell	JIP SINTIF Oil In Ice	2010	The spills were primarily aimed at improving understanding of the weathering and ignitability of spilled crude. However, they provided an opportunity for Shell staff supported under the JIP tto obtain the first ever measurements of methane emissions from such spills. Those emission rates are reported here. Gas dispersion modeling shows that the level of emissions from a significant spill are probably sufficient for its remote detection and mapping from a range of several kilometers using LightTouchˆ, Shell's hydrocarbon seepage detection technology. http://www.sintef.no/project/JIP_Oil_In_Ice/Dokumenter/publications/JIP-rep-no-23-MethaneFinalUnrestricted.pdf

(Continued)

Research Initiatives	Description	Facility	PI	Funding	Year	Objective
	New and Innovative Equipment and Technologies for the Remote Sensing and Surveillance of Oil in and Under Ice	CRREL	Dickins/ BSU	MMS, Statoil, ACS	2005	Project positively detected oil trapped in and under ice with two completely independent technologies
	Developing New and Innovative Equipment and Technologies for the Remote Sensing and Surveillance of Oil in and Under Ice - Phase 2		Dickins/ BSU	MMS, Statoil, ACS	2005	Follow up on work to develop technologies to detect oil located in or under ice. See 517 and 569

Research Initiatives	Description	Facility	PI	Funding	Year	Objective
	Development of a New Generation Laser Fluorosensor		EC-ESD	MMS, EC, USCG, Canadian Transportation Development Centre, and the Canadian Petroleum Association	2002	Develop the new laser fluorosensor technology for the detection of oil on water, ice, and shorelines.
	Detection and Tracking of Oil Under Ice		Dickins	MMS	2002	Establish a baseline of information that can be used to plan field trials with prototype systems to detect and map oil under ice
Surface / Dogs	Using dogs to detect oil hidden in snow and ice–results from field training on Svalbard April 2008		SINTEF & Trondheim Hundeskole	JIP SINTIF Oil In Ice	2009	The main objective of this project has been to train dogs to find oil spills hidden in snow or ice. Previous tests performed during 2007 in a laboratory environment in Trondheim showed that dogs are able to detect and identify the smell of oil, both weathered crude and bunker fuels.

(Continued)

Research Initiatives	Description	Facility	PI	Funding	Year	Objective
						Outdoor tests in the Trondheim area have also shown that dogs detect the smell of oil and can find point source at an outdoor temperature down to -5C. http://www.sintef.no/project/JIP_Oil_In_Ice/Dokumenter/publications/JIP-rep-no-14-Oildog-snow-ice.pdf
	Oil Detection By Specially Trained Dogs				2007	Brandvik, P.J. and T. Buvik. Nov 2007. *Oil Detection By Specially Trained Dogs*. Report from Phase 1 feasibility study. Memo prepared by SINTEF and Trondheim Hundeskole as part of Oil-in-Ice JIP Project 5 – Remote Sensing
Vessel	Development of an Airborne Oil Spill Thickness Sensor		EC-ESD	MMS	2000	Develop the technology of oil slick thickness sensor for the measurement of oil slick thickness on water from an airborne platform

Research Initiatives	Description	Facility	PI	Funding	Year	Objective
Research Initiatives	Description	Facility	PI	Funding	Year	Objective
	Development of Improved Oil Spill Remote Sensing Techniques		EC-ESD	MMS & EC	1990	Three approaches were taken: Use of shipborne radar as a practical aid to spill control Study Infrared oil spill remote sensing Study dispersant signature in remote-sensing imagery.
	Shipboard Navigational Radar as an Oil Spill Tracking Tool		MMS	MMS	1989	Refine the use of existing shipboard navigational radar units to detect and track oil spill slicks
Airborne	Detecting Oil On and Under Sea Ice Using Ground Penetrating Radar: Development of a New Airborne System		Dickins, BSU	ACS, BSEE, Conoco Phillips, ExxonMobil, Shell, Statoil	2010-12	Expand the practical operating window for oil detection with ground penetrating radar (GPR) to cover a wider range of sea ice and climate conditions. This project is a direct continuation of TAR projects 348, 517, 547, 569, and 588

(Continued)

Research Initiatives	Description	Facility	PI	Funding	Year	Objective
	Evaluation of Airborne Remote Sensing Systems for Oil in Ice Detection		Dickins & Norconsult	JIP SINTIF Oil In Ice	2010	The airborne programs in 2008 and 2009 provided real-world demonstrations of the capabilities and limitations—technical and operational—of using airborne surveillance aircraft to detect and monitor spills confined by ice in remote Arctic areas. http://www.sintef.no/project/JIP_Oil_In_Ice/Dokumenter/publications/JIP-rep-no-28-AirborneRSFinal.pdf
	Airborne GPR to Detect Oil Under Snow		BSU, Dickins, SINTEF	JIP SINTIF Oil In Ice	2010	Numerical modeling shows that GPR is sensitive to the presence of oil in the snow pack over a broad range of snow densities and oil types. Results of a controlled field experiment using a helicopter borne, 1000 MHz GPR system, showed that a 2 cm thick oil film trapped between snow and sea ice

Research Initiatives	Description	Facility	PI	Funding	Year	Objective
						was detected based on a 51% decrease in reflection strength. Our results indicate that GPR has the potential to become a robust tool that can substantially improve oil spill characterization and remediation. http://www.sintef.no/project/JIP_Oil_In_Ice/Dokumenter/publications/JIP-rep-no-24-GPRFinal.pdf
	Open Water Multispectral Aerial Sensor Oil Spill Thickness Mapping In Arctic and High Sediment Load Conditions		Ocean Imaging Corporation	BSEE	2011	Temperate sea and atmospheric conditions with reasonable water clarity do not exist in the Arctic conditions at high latitudes. There is a need for the testing of simplified, self-contained multispectral system configurations. This project is a direct continuation of TAR projects 544 and 594.

(Continued)

Research Initiatives	Description	Facility	PI	Funding	Year	Objective
	Development of a Portable Multispectral Aerial Sensor for Real- Time Oil Spill Thickness Mapping in Coastal and Offshore Waters	Ohmsett	Ocean Imaging Corporation	MMS & Cala F&G- Oil Spill Prevention and Response	2009	Develop a portable, easy-to-operate aerial sensor to detect and accurately map the thickness and distribution of an oil slick in coastal and offshore waters in real time
	Detection of Oil on and Under Ice - Phase 3		Dickins & BSU	MMS	2008	The development of oil and ice detection system has made significant progress through a series of successful projects (TAR-517, TAR-547 and TAR-569). This new research project will undertake a series of four tasks to assess the technical feasibility and cost of developing and incorporating airborne oil detection systems in future field trials with oil and ice.

Research Initiatives	Description	Facility	PI	Funding	Year	Objective
	Real-time Detection of Oil Slick Thickness Patterns with a Portable Multispectral Sensor		Ocean Imaging Corporation	MMS	2007	Develop an oil slick thickness measurement algorithm with laboratory and field validation.
	Development of a Frequency Scanning Radiometer to Measure Oil Slick Thickness, Phase II		USAF / MIT, Lincoln Laboratory	USAF	2000	Development of an all-weather, airborne instrument capable of measuring oil thickness over an oil-slick area in real time.
Satellite	Satellite Tracked Drifter Measurements in the NE Chukchi Sea			BOEM-AK	ongoing	Chukchi Sea
	Detection of Oil Under Ice Using Satellite	CRREL	TAS		2012	Test ability to detect oil under ice leveraged with TAR 659.

(Continued)

Research Initiatives	Description	Facility	PI	Funding	Year	Objective
	The Utilization of Satellite Images for the Oil in Ice Experiment in the Barents Sea. May 2009		NERSC & KSAT	JIP SINTIF Oil In Ice	2010	Twenty-six satellite images were received for analysis as a part of the Oil in Ice project: 10 Envisat ASAR wide swath images, 6 Radarsat-l images, 7 Radarsat-2 images, and 3 COSMOôSkyMed. The main objectives were to determine which images cover the activities, and what can be seen in these images with regards to ships, oil, and ship tracks through ice. http://www.sintef.no/project/JIP_Oil_In_Ice/Dokumenter/publications/JIP-rep-no-29-SatelliteFinal.pdf
	Evaluation of Satellite Data in Determining Pollution Resulting from Damage to the Oil & Gas		Naval Research Lab	MMS	2008	Identifying damage to the pipelines and structures in the Gulf of Mexico and resulting oil pollution by sensors on board satellites such as the such as the Advanced Very High Resolution Radiometer

Research Initiatives	Description	Facility	PI	Funding	Year	Objective
	Industry Infrastructure of the GOM Project Canceled					(AVHRR) on board the NOAA-n satellite series, and the Moderate-resolution Imaging Spectroradiometer (MODIS) on board of the TERRA and AQUA satellites.
	Using Satellite Radar Imagery to Detect Leaking Abandoned Wells on the US Oouter Continental Shelf		Advanced Resources Internationa l	MMS	2004	Use satellite radar imagery to detect if there are leaking abandoned wells.
Numerical Models	Review of Oil Spill Trajectory Modelling in the Presence of Ice		COOGER	COOGER	2011	The review addresses all the components of a comprehensive oil spill trajectory model, including: (1) a blowout plume model to determine the distribution of oil in the water column for spills that occur at depth, (2) models for the physical environmental forcing (wind, air temperature, precipitation, ocean currents, sea ice, and waves), and

(Continued)

Research Initiatives	Description	Facility	PI	Funding	Year	Objective
						(3) an oil fate-and-effects model to address weathering, evaporation, ice-oil interactions, and other details of the oil's interplay with the environment. http://www.dfo-mpo.gc.ca/Library/344774.pdf
	OWM Weathering model		SINTEF		2011	(OWM) provides comprehensive predictions of the behavior of spilled oil at sea under different weather conditions. Excellent agreement with field observations has been consistently achieved with this model when predictions are based on standardized laboratory studies of specific oils and petroleum products
	Modelling of Oil in Ice with OSCAR		SINTEF	JIP SINTEF Oil In Ice	2011	Fresh crude oil was released uncontained between the ice floes to study oil weathering and spreading in the ice by multiple

Research Initiatives	Description	Facility	PI	Funding	Year	Objective
						sampling throughout the six-day experiment, including meteorological and oceanographic (MetOcean) data. The objective of the comprehensive sampling program was to acquire more knowledge of how the presence of ice influences the distribution and spreading of oil on the surface and in the water column. http://www.sintef.no/project/JIP_Oil_In_Ice/Dokumenter/publications/A19804%20JIP-report-no%2035%20Petromaks%20modelling%20Final.pdf
	OSCAR Model		SINTEF		2011	Develop models addressing effects and risks associated with operational and accidental discharges to the marine environment. The OSCAR model provides a scenario-based and 3-D statistic approach to objective evaluation of alternative response strategies.

(Continued)

Research Initiatives	Description	Facility	PI	Funding	Year	Objective
						Both physical and biological measures of success and effectiveness are provided. This 3-D map-based system uses the SINTEF OWM for weathering calculations.
	Experiment Oil Release in Broken Ice – A Large-Scale Filed Verificatioon of Results From Laboratory Studies of Oil Weathering and Ignitability of Weathered Oil Spills		SINTEF	JIP SINTEF Oil In Ice	2010	This field data from the large-scale experiment (70–90% ice coverage) correspond well with data from previous mesoscale basin experiments at similar ice coverage. This confirms that the results from the weathering experiments performed in the mesoscale basins at SINTEF SeaLab can be interpreted as reflecting realistic conditions. This is a significant verification since the possibility to perform large-scale field experiments is limited.

Research Initiatives	Description	Facility	PI	Funding	Year	Objective
						The data set from the basin studies and field verification is used to improve SINTEF Oil Weathering Model's capability to predict weathering of oil spills: in cold conditions and ice. http://www.sintef.no/project/JIP_Oil_In_Ice/Dokumenter/publications/JIP-rep-no-26-FEX2009-weathering-ISB%20final.pdf
Mitigation						
Planned	Maximum Credible Blowout Occurance and Size Estimators for AK OCS		NA	BOEM-AK	2012	Beaufort, Chukchi
	AK Environmental Database		NA	BOEM-AK	2012	Beaufort, Chukchi
	Characterization of the Circulation on the		NA	BOEM-AK	2012	Beaufort, Chukchi

(Continued)

Research Initiatives	Description	Facility	PI	Funding	Year	Objective
	Continental Shelf Areas of the Northeast Chukchi and Western Beaufort Seas					
	Applications for Mapping Spilled Oil in Arctic Waters		NA	BOEM-AK	2012	Chukchi
	Adaption of Arctic Circulation Model		NA	BOEM-AK	2011	Chukchi
	Updates to the Fault Tree for Oil-Spill Occurance Estimators Needed Under Forthcomming BOEM 5 year program		NA	BOEM-AK	2011	Beaufort, Chukchi

Research Initiatives	Description	Facility	PI	Funding	Year	Objective
On Going	Oil Spill Occurrence Estimator for Onshore AK North Slope Crude and Refined Oil Spills Training and Response (STAR) Calculator Program		NA	BOEM-AK	2011	Beaufort, Chukchi
	Shore Zone Shoreline Mapping of the North Slope AK		NA	BOEM-AK	2011	Beaufort, Chukchi
Response	Ice Regimes for Oil Spill Response Planning		SINTEF, Oil Spill Consultant, Counterspil Research	JIP SINTIF Oil In Ice	2008	The ice conditions were reviewed for nineteen regions where oil exploration and production (E&P) activities are either ongoing or planned. Using internationally accepted classification terminology to describe the ice, the immediate

(Continued)

Research Initiatives	Description	Facility	PI	Funding	Year	Objective
						objective of the work became one of identifying ice regimes common to most circumpolar areas of the world. There are some overall similarities in the ice conditions of the 19 regions that have permitted grouping according to five ice regimes: i.e. Arctic shallow semi-enclosed sea, Arctic open-sea, Arctic and sub-Arctic coastal, sub-Arctic estuary and non-Arctic shallow sea. http://www.sintef.no/project/JIP_Oil_In_Ice/Dokumenter/publications/JIP-rep-no-15-Ice%20regimes-final_2010.pdf
	Combining Mineral Fines with Chemical Dispersants to Disperse Oil in		DoFO	BSEE	2012	Assess the feasibility of a cold water and Arctic marine oil spill countermeasure strategy based on the stimulation of oil-mineral aggregate (OMA)

Research Initiatives	Description	Facility	PI	Funding	Year	Objective
	Low Temperature and Low Mixing Energy Environments					formation in the presence of a chemical dispersant.
	Research on Improving Methods for Recovering Residues from In Situ Burning of Marine Oil Spills		SL Ross	BSEE	2011	Research program is to develop methods for (1) recovering potentially nonbuoyant ISB residues through innovative surface and subsurface collection mean, and (2) recovering buoyant residues, including those attached to sorbent agents that were intentionally added to the residue to prevent subsequent residue sinking.
	Effective Daily Recovery Capacity (EDRC) Project		Genwest Systems, Inc.	BSEE	2011	Deepwater Horizon oil spill response has highlighted that the EDRC for an oil skimmer may not be an effective or accurate planning standard and predictor of oil response equipment recovery capacity. EDRC is a regulation that greatly influences vessel and facility response planning standards

(Continued)

Research Initiatives	Description	Facility	PI	Funding	Year	Objective
						nationwide and established the basis for the initial capitalization of Oil Spill Removal Organizations (OSROs).
	Oil Spill Training and Response (STAR) Calculator Program		Genwest Systems, Inc.	MMS, Shell International Exploration and Production, Inc. and the American Petroleum Institute. NOAA	2010	Project are to standardize the existing software packages; to enhance their utility, user-interface and output; and to integrate all three response options (mechanical, burning, and dispersants) using improved algorithms for their efficient use under a variety of spill scenarios.
	Mitigating Oil Spills from Offshore Oil and Gas Activities by Enhancement of Oil-Mineral Aggregate Formation		COOGER	MMS & Department of Fisheries and Oceans	2009	Project will assess the feasibility of a marine oil spill countermeasure strategy based on the stimulation of OMA formation. Experiments will be conducted on both laboratory and wave tank systems under controlled conditions to evaluate its potential effectiveness for the

Research Initiatives	Description	Facility	PI	Funding	Year	Objective
						treatment of oil spills from ships, facilities, or pipelines. http://www.dfo-mpo.gc.ca/science/publications/article/2009/05- 19-09-eng.htm
	Oil Spill Management		SINTEF		2005	Oils cause them to behave very differently when spilled at sea. Also, recent incidents of various light, medium, and heavy bunker fuel oils like the North Cape Spill in the US, Erika incident in France, the Baltic Carrier in Denmark, and the Green Ålesund and John R incidents in Norway have shown a wide spectre in physical behavior, biological effects, and in the performance of response options. Prior knowledge of the likely behavior and weathering of various oils and pre-spill analyses of the feasibility and effectiveness of different response strategies under various environmental conditions,

(Continued)

Research Initiatives	Description	Facility	PI	Funding	Year	Objective
						are essential part of any oil spill contingency planning to optimize the overall "Net Environmental Benefit" of a combat operation.
	Production of a White Paper and Workshop Regarding a Full Scale Experimental Oil Release in the Barents Sea Marginal Ice Zone		SINTEF	MMS	2004	Evaluated alternative oil spill response options and in supplying data strengthened model simulations of oil-ice interactions.
	Mechanical Oil Recovery in Ice Infested Waters (MORICE) - Phase III	OHMSETT	SINTEF	MMS, Alaska Clean Seas, Prince William Sound Oil Spill Recovery Institute, BP	2003	Develop technologies for more effective recovery of oil spills in ice-infested waters.

Research Initiatives	Description	Facility	PI	Funding	Year	Objective
				Exploration, Phillips Alaska, Norsk Hydro and Store Norske Spitzbergen Kulkompani		
	A Method to Determine Worst Case Discharges From Facilities That Produce Or Transport Oil in the U.S. Outer Continental Shelf (OCS)		SINTEF	MMS and 23 different oil companies to conduct this research	2003	Produced a model to predict a discharge from a pipeline. It also included a pocket guide to quickly make an estimate of a worst case discharge from a pipeline. The model is known as the MMS Pipeline Oil Spill Volume Estimation Model (POSVEM). POSVEM is a computer-based methodology to estimate discharges from seafloor pipelines.
	Comprehensive Spill Response Tactics for the Alaska North Slope-Oil in		SL Ross, DF Dickins Asscociates, Vaudrey	MMS and Alaska Clean Seas	1998	Developed comprehensive oil spill response plans for petroleum operations on the North Slope of Alaska. The objective of the study was to evaluate the capabilities to

(Continued)

Research Initiatives	Description	Facility	PI	Funding	Year	Objective
	Broken Ice Spill Response Scenarios					recover spilled oil from very large oil well blowouts occurring during broken ice conditions in the southern Beaufort Sea.
	World Catalog of Oil Spill Response Products		Environ-mental Consultants Inc.	MMS, USCG, EC	1991	Equipment listed in catalog has become obsolete. Updated list available at http://www.slross.com/WorldCat/WorldCatmain.htm
Containment Booms	Fighting Oil Spills with Air Bubbles		SINTEF		2009	Obvious advantages in using the bubble curtain instead of traditional oil-booms: it is an efficient way of closing off a vulnerable area to prevent an oil-spill from entering it. It also enables us to limit the spread of a spill and improves our prospects of collecting the oil.
	Oil Spill Response - Performance Review of Booms		Environ-mental Consultant	MMS	2005	Analyzed and summarized all available performance test data of oil spill containment boom.

Research Initiatives	Description	Facility	PI	Funding	Year	Objective
	The Use of Ice Booms for the Recovery of Oil Spills from Ice Infested Waters		Fleet Technology Limited	MMS	2004	The objective of this research contract was to study the technology in the design and use of ice booms for recovering spilled oil in ice infested waters.
	Effect of Oil Spill Containment Boom Characteristics on Boom Performance	Ohmsett	SL Ross	MMS	2004	Research investigated the effect of a previously identified key containment boom characteristic on boom performance.
	Estimation of Towing Forces on Oil Spill Containment Booms	Ohmsett		MMS	2001	The purpose of the study is to determine the maximum towing forces allowable for different boom types.
	Numerical Modeling of Oil Boom Behavior and Rapid Current Boom Development		URI, UNH	USCG	2000	Develop a rapid current boom using the submergence plane concept.

(Continued)

Research Initiatives	Description	Facility	PI	Funding	Year	Objective
	Field Experiments at the Ohmsett Facility, Especially for a Newly Designed Boom System	Ohmsett	University of Miami	MMS	2000	Developed a new boom design that uses inclined plane technology project was to build a full-scale prototype and test it at Ohmsett.
	Water Jet Barrier Containment of Oil in the Presence of Broken Ice		EC-ESD	MMS and EC	1997	Optimize the design of an innovative oil spill containment barrier. The barrier uses high pressure water and fine droplets above and parallel to the water surface resulting in a high-velocity local wind.
Herders	Chemical Herders to Expand ISB Window of Opportunity		OGP	JIP- AOSRT	2012	Conduct meso- and large-scale basin research and field verification experiments with chemical herders to enhance and improve the effectiveness of ISB in specific Arctic ice environments.

Research Initiatives	Description	Facility	PI	Funding	Year	Objective
	Field Testing of the USN Oil Herding Agend on Heidrun Crude in Loose Drift Ice		SINTEF & SL Ross	JIP SINTIF Oil In Ice	2010	A two-day field research program was conducted off Svalbard in late May 2008 to test the efficacy of a chemical herding agent in thickening oil slicks on water among very open drift ice for subsequent in situ burning. http://www.sintef.no/project/JIP_Oil_In_Ice/Dokumenter/publications/JIP-rep-no-6-FEX2008-Herders-Final.pdf
	Employing Chemical Herders to Improve Oil Spill Response Operations	Ohmsett	SL Ross	MMS	2011	Extend the research on chemical herders into pack ice conditions (mechanical containment recovery), in salt marshes (mechanical recovery, and in situ burning), and in open water (dispersants). This project was a direct continuation of TAR Project 554 Mid-Scale Test Tank Research on Using Oil Herding Surfactants to Thicken Oil Slicks in Broken Ice.

(Continued)

Research Initiatives	Description	Facility	PI	Funding	Year	Objective
						As a skimmer removes oil from the center of a herded slick, the action of the herding agent may cause the slick to continuously contract towards the skimmer, eliminating the need to move the skimmer around to contact all the oil.
	Mid-Scale Test Tank Research On Using Oil Herding Surfactants To Thicken Oil Slicks In Broken Ice	CRREL	SL Ross	MMS,ExxonMobil &PERF	2007	Research on the use of chemical herding agents to thicken oil spills in broken ice to allow them to be effectively ignited and burned in situ.
In-Situ Burning (ISB) Fire Booms						

Research Initiatives	Description	Facility	PI	Funding	Year	Objective
	State of Knowledge		OGP	JIP-AOSRT	2013	To prepare educational and outreach materials to educate and make stakeholders aware of the significant body of existing knowledge on ISB.
	Aerial Ignition Systems		OGP	JIP-AOSRT	2013	To provide technology improvement that will deliver a reliable means of aerial ignition and improve oil slick targeting to support the use of in situ burning.
	Using Oil Herding Agents for Rapid Response in situ Burning of Oil Slicks on Open Water	Ohmsett	SL Ross	BSEE	2012	Evaluate the feasibility of using herders to enable in situ burning as a rapid-response technique in open water. This research was accomplished by performing experiments in the laboratory with the US Navy's hydrocarbon based herder formulation and the best silicone herder formulation to find the most effective product for various water temperatures. Continuation of TAR Projects 554 and 617

(Continued)

Research Initiatives	Description	Facility	PI	Funding	Year	Objective
	Tests of Fire-Resistant Booms in Low Concentrations of Drift Ice- Field Experiment May 2009		SL Ross	JIP SINTIF Oil In Ice	2010	The goal in the 2009 tests was to tow the same two sections of fire-resistant boom through a field of drift ice containing oil, and then burn the oil in situ. Boom ås mechanical performance was tested in 2008. http://www.sintef.no/project/JIP_Oil_In_Ice/Dokumenter/publications/JIP-rep-no-27-FEX2009-FireboomFieldTests-final.pdf
	Experiment Oil Release in Broken Ice- A Large-Scale Filed Verificatioon of Results From Laboratory Studies of Oil Weathering and		NA	JIP SINTIF Oil In Ice	2010	Ignitability for In Situ Burning of Oil Spills. For removal of oil spilled in ice-infested waters, in situ burning (ISS) is one of the response techniques with the highest potential for Arctic conditions especially in snow and dense ice.

Research Initiatives	Description	Facility	PI	Funding	Year	Objective
	Ignitability of Weathered Oil Spills					In order to make in situ burning an operational tool, there is at need to better define the potential and limitations for ignition of oil spills with regard to oil type and weathering degree. http://www.sintef.no/project/JIP_Oil_In_Ice/Dokumenter/publications/JIP-rep-no-26-FEX2009-weathering-ISB%20final.pdf
	Establishing, Testing, and Verification of a Laboratory Burning Cell to Measure Ignitability for In Situ Burning of Oil Spills		SINTEF	JIP SINTIF Oil In Ice	2010	For removal of oil spilled in ice-infested waters, in situ burning is one of the response techniques with the highest potential for Arctic conditions especially in snow and dense ice. In order to make In situ burning an operational tool, there is at need to better define the potential and limitations for ignition of oil spills with regard to oil type and weathering degree. http://www. sintef.no/project/JIP_Oil_In_Ice/Dokumenter/publica tions/JIP-rep-no-20%20Burning-cell-final.pdf

(Continued)

Research Initiatives	Description	Facility	PI	Funding	Year	Objective
	Tests of Fire-Resistant Booms in Low Concentrations of Drift Ice – Field Experiment May 2008		SINTEF & SL Ross	JIP SINTIF Oil In Ice	2010	The overall objective of this portion of the field program was to determine whether fire-resistant containment boom can be used to aid in burning oil in situ in lower drift ice concentrations. http://www.sintef.no/project/JIP_Oil_In_Ice/Dokumenter/publications/JIP-rep-no-5-FEX2008-FireboomFieldTests-final.pdf
	Project Cancelled: In Situ Oil Burning Manual Preparation		EC	MMS, EC, CCG, USCG	2010	Research on the environmental effects of burning oil in situ.
	Analysis of Oil-Slick Combustion		NIST	MMS	2005	Focuses on the burning characteristics of crude oil fires on water; ALOFT--FT (A Large Outdoor Fire plume Trajectory model).http://fire.nist.gov/aloft/.

Research Initiatives	Description	Facility	PI	Funding	Year	Objective
	Fire Boom Testing at Ohmsett	Ohmsett	SL Ross	MMS	2004	fire exposure portion of the test involves three cycles of one-hour exposure to air-enhanced propane flames in waves, followed by a one hour cool down period in waves alone,
	Mid-Scale Tests to Determine the Limits to In Situ Burning in Broken Ice		SL Ross	MMS	2004	This purpose of this research project is to investigate the minimum ignitable thickness, combustion rate, effects of waves, and residue amount for crude oils burned in situ in cold water and broken ice conditions.
	Laboratory Testing to Determine In Situ Burning Parameters for Six Additional U.S. OCS Crude Oil		SL Ross	MMS	2000	Six additional oils from OCS facilities are being analyzed to determine if their suitability of spill response by in situ burn methods. See 288.

(Continued)

Research Initiatives	Description	Facility	PI	Funding	Year	Objective
	Testing of Fire Resistant Booms in Waves and Flames		SL Ross	MMS & CCG	1999	Develop a near full-scale screening test protocol for the effectiveness and durability of fire resistant oil containment boom that incorporates simultaneous testing in waves and flames.
	Laboratory Testing to Determine Operational Parameters for In Situ Burning of U.S. OCS Crude Oil Spills		SL Ross	MMS	1999	Baseline burn tests were conducted to determine the natural burning characteristics of water-free and emulsified slicks of fresh and weathered oils. Burn efficiency and burn rate were calculated from the data gathered in the burn tests.
	Re-Engineering of a Stainless Steel Fireproof Boom for Using in Conjunction with Conventional Firebooms	Ohmsett	SL Ross	MMS	1999	Re-engineer an existing stainless steel fire boom. The new boom design is reduced in size, weight, and cost.

Research Initiatives	Description	Facility	PI	Funding	Year	Objective
	Technology Assessment and Concept Evaluation for Alternative Approaches to In Situ Burning of Oil in the Marine Environment		Marine Research Associates LLC	MMS	1998	Assess the technical and operational feasibility of two alternative approaches to in situ burning of oil spills.
	Outdoor Wave Tank and Program of Mid-Scale In Situ Burn Testing in Alaska		SL Ross	Alaska Department of Environmental Conservation, Alaska Clean Seas, BP Exploration	1997	Consisted of a modest laboratory-scale program to evaluate the burning characteristics of four Alaska risk oils.
	Recovery Methods for High Viscosity Oils		EC-ESD	MMS & EC	1991	Evaluate two different atomization systems to be used in a catamaran-type burner to recover high viscosity oils.
	Helicopter-Borne Laser Ignition of Oil Spills		Physical Sciences Inc.	MMS& EC	1990	Evaluate necessary components for the development of airborne laser ignition of oil spilled on the open ocean.

(Continued)

Research Initiatives	Description	Facility	PI	Funding	Year	Objective
Dispersants	Fate of Dispersed Oil Under Dynamic Drift and Pack Ice		OGP	JIP-AOSRT	2013	Develop a numerical model to predict the fate of a dispersed oil plume that develops under ice, particularly the resurfacing potential.
	Dispersant Testing Under Realistic Field Conditions		OGP	JIP-AOSRT	2013	Understand operational needs for dispersant and mineral fines application in Arctic conditions. Conduct large-scale basin tests and field verification on the efficacy of dispersant and mineral fines in Arctic marine waters.
	Validation of the Two Models Developed to Predict the Window of Opportunity for Dispersant Use in the Gulf of Mexico		EC	BSEE	2012	BSEE-funded research project entitled: Identification of Window of Opportunity for Chemical Dispersants on Gulf of Mexico Crude Oils. Two correlation models were developed to predict the window of opportunity (or time window) for successful chemical

Research Initiatives	Description	Facility	PI	Funding	Year	Objective
						dispersant use in the Gulf of Mexico. The models consist of correlation relationships established using best-fit correlation between readily available fresh oil properties and the window of opportunity for successful chemical dispersant.
	Chemical Dispersant Research at Ohmsett: Phase 2		SL Ross	BSEE	2011	A review of oil spill dispersants, their efficacy and effects, recently completed by the US National Research Council (NRC 2005), recommended that research on chemical dispersants be conducted in several different areas. See TAR project 615.
	Baffled Flask Dispersant Effectiveness Testing		EPA	BSEE	2011	Large-scale test tank produced DE data on 20 crude and fuel oils using Corexit 9500 dispersant. Laboratory-scale DE tests at Ohmsett will be conducted using the same dispersant and oil combinations, and results will be compared to determine if the

(Continued)

Research Initiatives	Description	Facility	PI	Funding	Year	Objective
						bench-scale test results can be used to provide reasonable estimates of field performance.
	The effects of use on dispersants and in situ burning on Arctic marine organisms- A laboratory Study		SINTEF	JIP SINTIF Oil In Ice	2011	The effect and toxicity of a water soluble fraction (WSF) of oil versus oil added dispersants, and WSF versus the underling water after in situ burn, were studied. The exposure concentrations used were based on monitoring of the WSF in the water column during an offshore field experiment. http://www.sintef.no/project/JIP_Oil_In_Ice/Dokumenter/publications/A19803%20Jip-rep-no-34-Petromaks%20exposurefinal.pdf

Research Initiatives	Description	Facility	PI	Funding	Year	Objective
	Laboratory-Scale Investigation of a Method for Enhancing the Effectiveness of Oil Dispersants in Destabilizing Water-in-Oil Emulsions		Auburn University	BSEE	2011	Investigated the feasibility of enhancing the de-emulsifying properties of commercially available oil dispersants by modifying the composition and fraction of polar constituents in the oil phase of water-in-oil emulsions and increasing the pH of the emulsion aqueous phase.
	Operational Chemical Dispersant Research at Ohmsett	Ohmsett	SL Ross	BSEE	2011	(DWH) oil spill and subsequent use of large quantities of chemical dispersants in the response to the spill has heightened the interest in the use of dispersants questions concerning the use of dispersants so regulators, decision makers, end-users, and the public can make informed decisions regarding their use.This proposed project is a direct continuation of TAR Project 638 "Chemical Dispersant Research at Ohmsett: Phase 2."

(Continued)

Research Initiatives	Description	Facility	PI	Funding	Year	Objective
	Assessment of Dispersant Effectiveness using Ultrasound to Measure Oil Droplet Particle Size Distributions		Applied Research Associates, Inc. (ARA)	BSEE	2011	Develop novel ultrasonic scattering methods to measure the droplet size of dispersed oil to provide technologies to monitor the efficacy of dispersants subsea. ARA will develop ultrasonic measurements to determine the crude oil droplet size of dispersed oil to monitor/determine the efficacy of dispersants as a function of oil type, dispersant type, dispersant-to-oil ratio, water temperature, oil temperatures, and the presence of sediment on the effectiveness of dispersants.
	Development and testing of a containerized dispersant spray system for use in cold and ice-covered area		SINTEF	JIP SINTIF Oil In Ice	2010	The overall goal has been to optimize and improve the methodology and strategies for dispersant response operations in cold and ice-covered areas. This paper summarizes the systematic

Research Initiatives	Description	Facility	PI	Funding	Year	Objective
						and scientific approach in this technological development and tests documentation of the containerized dispersant spray system. http://www.sintef.no/project/JIP_Oil_In_Ice/Dokumenter/publications/JIP-rep-no-13-Development-of-spray-arm-final.pdf
	Evaluation of Dispersant Spray Systems and Platforms for Use on Spilled Oil in Seas with Ice Present (JIP Project 4, Act 4.21)		SINTEF	JIP SINTIF Oil In Ice	2007	This report gives a short overview and evaluations of different dispersant application platforms for use in ice-covered areas. http://www.sintef.no/project/JIP_Oil_In_Ice/Dokumenter/publications/JIP-rep-no-12-Evaluation-of-dispersants-spray-platforms-200207.pdf
	A Review of Studies of Oil Spill Dispersant Effectiveness in		SINTEF	JIP SINTIF Oil In Ice	2007	This report gives a short state-of-the art of the testing oil spill dispersants in simulated cold conditions that has been undertaken by several research groups, but with

(Continued)

Research Initiatives	Description	Facility	PI	Funding	Year	Objective
	Arctic Conditions. (JIP Project 4, Act.4.11)					different combinations of low temperature, absence or presence of ice, and water salinity. This report was to ensure transfer of newest knowledge and a good planning and experimental design of the dispersant effectiveness testing in task 4.1. http://www.sintef.no/project/JIP_Oil_In_Ice/Dokumenter/publications/JIP-rep-no-11-Dispersant-Effectiveness-in-ArcticConditions-150207.pdf
	Chemical Dispersant Research at Ohmsett	Ohmsett	SL Ross	MMS	2010	PERF funded a research project conducted by SINTEF in Norway and Cedre in France that has looked at the same issue of dispersant wash out but at laboratory bench scale. The Oseberg crude oil used in the PERF study was shipped to Ohmsett and was used in the 2007 study (TAR Project 563).

Research Initiatives	Description	Facility	PI	Funding	Year	Objective
	Literature Review on Chemical Treating Agents in Fresh and Brackish Water		SL Ross	MMS	2010	Chemical dispersants are designed to work effectively in salt water (35 ppt salinity). Nearshore environments water in these areas will be fresh (0% salinity) and brackish (10–15% salinity) and this may alter the effectiveness of chemical treating agents and dispersants and thus alter the treating agents and dispersant use decision. The three regional representatives on the MMS oil spill response research team identified an information gap and requested a comprehensive review of the of the effectiveness of chemical treating agents, including dispersants in fresh and brackish water.
	Heavy Oil Dispersion Research		SL Ross	BSEE	2010	The objective of the project was to study the mechanism by which oil viscosity limits the effectiveness of dispersants. Specifically, two viscosity issues were studied.

(Continued)

Research Initiatives	Description	Facility	PI	Funding	Year	Objective
						One was the ability of the dispersant to penetrate into viscous oil upon initial application prior to being washed away by surface water. The other was the internal visco-elasticity of the oil-dispersant mix (in conjunction with the dispersant dosage that has successfully mixed into the oil) that may prevent the oil from being broken into droplets when wave energy is applied.
	Development of a Training Package on the Use of Chemical Dispersants for Ohmsett – The National Oil Spill Response Test Facility	Ohmsett	SL Ross	MMS	2009	This project will develop a chemical dispersant training course to be conducted at Ohmsett that includes practical hands-on experience with handling, safety, application, monitoring, efficacy, and recovery in breaking wave environments.

Research Initiatives	Description	Facility	PI	Funding	Year	Objective
	Evaluating the Biodegradability and Effects of Dispersed Oil using Arctic Test Species and Conditions		UAF, NewFields Northwest	JIP with Shell, ExxonMobil, Statoil and ConocoPhillips	2008	Addressed the following data gaps: (1) the toxicity of physically and chemically dispersed oil on pelagic species that are key components of Arctic food webs, and (2) the biodegradation of physically and chemically dispersed oil released into pelagic waters using indigenous Arctic microbes under Arctic conditions. see AMOP 2011 proceeding
	Understanding the Effects of Time and Energy on the Effectiveness of Dispersants		SINTEF	MMS	2008	Designed to gather data to support decision makers in the process of determining whether dispersants should be used in low energy environments. This information will be useful for dispersant decision making in ice cover (an ice field reduces wave motion) or other calm conditions.

(Continued)

Research Initiatives	Description	Facility	PI	Funding	Year	Objective
	Investigation of the Ability to Effectively Recover Oil Following Dispersant Application	Ohmsett	SL Ross	MMS	2008	Determine whether the application of dispersants to an oil slick reduces the ability to subsequently recover oil with conventional skimming systems.
	Changes with Dispersant Effectiveness with Extended Exposure in Calm Seas	Ohmsett	SL Ross & Alun Lewis Oil Spill Consultancy		2008	The project is to continue to investigate the conditions that might lead to the loss of surfactants from dispersant-treated oil so that subsequent application of breaking waves will not result in dispersion.
	Upgrade of SMART Dispersant Effectiveness Monitoring Protocol		SL Ross	MMS	2008	Analysis of monitoring data (visual and instrumental monitoring) collected during Ohmsett dispersant experiments, completed between 2003 through 2007, for the purposes of verifying the reliability of existing SMART effectiveness monitoring protocols.

Research Initiatives	Description	Facility	PI	Funding	Year	Objective
	Dispersant Effectiveness Testing on Realistic Emulsions at Ohmsett	Ohmsett	SL Ross	MMS	2007	Determine the viscosity limit for the effective application of chemical dispersants to realistic emulsions. Large-scale dispersant effectiveness experiments will be conducted at Ohmsett on these emulsions using Corexit 9500 and Corexit 9527 dispersants.
	Calm Sea Application of Dispersants	Ohmsett	SL Ross	MMS, ExxonMobil, & PERF	2007	Determine the period of time that oil spill dispersants applied to spilled oil in a calm sea will remain effective before the sea state increases and dispersion occurs.
	Dispersant Effectiveness Test Protocol Development for Ohmsett	Ohmsett	SL Ross	MMS	2006	Developed a standard test protocol for a wide range of dispersant related variables at the Ohmsett facility.
	Correlating Results of Dispersants Effectiveness at	Ohmsett	SL Ross	MMS	2006	Determine if the results of dispersant effectiveness tests are consistent with laboratory and actual at-sea conditions deployment of the Surveillance and Monitoring

(Continued)

Research Initiatives	Description	Facility	PI	Funding	Year	Objective
	Ohmsett with Identical At-Sea Trials: Effects of Oil Viscosity and Dispersant to Oil Ratio					for Alternative Response Technology monitoring protocols in experiments where direct, independent measurements of effectiveness.
	Dispersant Effectiveness Testing on Heavy OCS Crude Oils at Ohmsett		SL Ross	MMS	2006	Project will determine the limiting viscosity for the effectiveness of chemical dispersants applied to viscous US Outer Continental Shelf (OCS) crude oils from the Gulf of Mexico and Pacific OCS.
	Correlate Ohmsett Dispersant Tests with At-Sea Trials; Supplemental Tests to Complete Test Matrix	Ohmsett	SL Ross	MMS	2006	Testing will be conducted on IFO-180 fuel oil treated with Corexit 9500 dispersant at a DOR of 1:25. In replicated tests at sea this combination yielded consistently high levels of dispersant performance. See TAR 477.

Research Initiatives	Description	Facility	PI	Funding	Year	Objective
	The Effect of Warming Viscous Oils Prior to Discharge on Dispersant Performance	Ohmsett	SL Ross	MMS	2006	Assess the dispersibility of fresh and weathered Alaskan and Canadian crude oils in very cold water. Results from these experiments indicate that the crude oils tested were dispersible at near freezing water temperatures.
	Analysis of Dispersant Effectiveness of Heavy Fuel Oils and Weathered Crude Oils at Two Different Temperatures Using the Baffled Flask Test		EPA and University of Cincinnati	US EPA and MMS	2006	Conduct laboratory effectiveness testing using two test protocols the BFT and the and SWT with three dispersants (Corexit 9500, SuperDispersant 25 and Agma DR 379), two oils (IFO-180 and IFO-380) at one Dispersant to Oil Ratio (DOR 1:25) at one temperature (16ÀC). See TAR 513.
	Chemical Dispensability of OCS Crude Oils in Non-Breaking Waves,	Ohmsett	SL Ross	MMS, ExxonMobil, & PERF	2006	Determine whether chemically treated low-viscosity OCS crude oils disperse in a nonbreaking wave environmental and if so, to determine whether there is a limiting oil viscosity for chemical

(Continued)

Research Initiatives	Description	Facility	PI	Funding	Year	Objective
	Part 1: Determining the Limiting Oil Viscosity for Dispersion in Non- Breaking					dispersion for OCS crude oils in nonbreaking waves.
	Research at Ohmsett on the Effectiveness of Chemical Dispersants on Alaskan Oils in Cold Water	Ohmsett	SL Ross & Dickins	MMS	2006	Large-scale dispersant experiments in very cold water at Ohmsett using Alaska North Slope, Endicott, Northstar and Pt. McIntyre crude oils; Corexit 9500 and Corexit 9527 dispersants were used in the two test series.
	Development of Dispersant Test Protocol at Ohmsett	Ohmsett	SL Ross	MMS	2005	Examined various ways that dispersant testing might be accomplished at Ohmsett and evaluated the feasibility and costs.
	Effects of Chemically Dispersed and Biodegraded Oils		Plymouth Laborato-ries, Inc.	MMS	2005	Determined the effects of chemically dispersed and biodegraded oils.

Research Initiatives	Description	Facility	PI	Funding	Year	Objective
	Understanding Oil Spill Dispersants: Efficacy and Effects		National Research Council, Ocean Studies Board	MMS, NOAA, API, and USCG	2005	Review and evaluate existing information regarding the efficacy and effects of dispersants as an oil spill response technique.
	Chemical Response to Oil Spill: Ecological Effects Research Forum (CROSERF)		Ecosystem Manage-ment and Associates, Inc.	API, Exxon, Chevron, Marine Spill Response Corporation, state government agencies (AK, CA, FL, LA, TX, WA), federal government agencies (MMS, NOAA, EPA) Env Canada	2004	Coordinate and disseminate the research information on dispersants funded by private industry (API, Exxon, Chevron, Marine Spill Response Corporation), state government agencies (AK, CA, FL, LA, TX, WA), federal government agencies (MMS, NOAA, EPA,) and Environment Canada.

(Continued)

Research Initiatives	Description	Facility	PI	Funding	Year	Objective
	Laboratory Study to Compare the Effectiveness of Chemical Dispersants When Applied Dilute versus Neat		SL Ross	MMS	2004	Determine if the application of dispersants in a dilute form reduces their effectiveness when compared with neat application.
	Assessment of the Use of Dispersants on Marine Oil Spills in California		SL Ross	MMS	2004	Assessment of the operational and environmental factors associated with the use of chemical dispersants to treat oil spills in California Marine waters, with a view toward expediting dispersant use decision making and planning for such spills.
	Using Dispersants to Test and Evaluate the Effectiveness of Dispersants in Cold Water and Broken Ice		SL Ross	MMS	2004	Evaluate the effectiveness of Corexit 9500 and Corexit 9527 dispersants on Alaskan and Canadian crude oils.

Research Initiatives	Description	Facility	PI	Funding	Year	Objective
	Ohmsett 2003 Cold Water Dispersant Effectiveness Experiments	Ohmsett	SL Ross	MMS	2004	Fourteen large-scale dispersant effectiveness experiments test were completed at the Ohmsett facility with various combinations of oil type and dispersant-to-oil ratios (DORs).
	Analysis of IFO-180 and IFO-380 Oil Properties for Dispersant Window of Opportunity	Ohmsett	SL Ross	MMS	2004	Extend the work to complete property analyses on IFO-180 and IFO-380 fuel oils used in the June 2003 UK field trials and in the Ohmsett dispersant effectiveness experiments completed in the fall of 2003.
	Correlating Results of Ohmsett Dispersant Test with At-Sea Trials: Workshop to Coordinate Publications and Prioritize Follow-up Research	Ohmsett	SL Ross	MMS	2004	Project was to coordinate the publications of results of the 2003 MMS-sponsored "Correlating Results of Ohmsett Dispersant Tests with Identical At-Sea Trials."

(Continued)

Research Initiatives	Description	Facility	PI	Funding	Year	Objective
	Study of Oil Spill Chemical Treating Agents		EC-ESD	MMS & EC	2002	1. Develop tests for oil spill chemical treating agents and develop more effective and improved products 2. Test commercial products for efficacy and toxicity 3. Test oil spill dispersants to understand effectiveness and how they might be improved 4.Test new product ideas (i.e. dispersants: emulsion breakers, biodegradation agents, solidifiers)
	Biodegradation of Chemically Dispersed Oil an Ecosystem Approach		AEA Technology	MMS, USCG, and Alaska Department of Environmental Conservation	2002	Determine the fate and effects of chemically dispersed oil in the context of an ecosystem approach.

Research Initiatives	Description	Facility	PI	Funding	Year	Objective
Mechnical Recovery	Mechanical Recovery in Ice-Infested Waters		OGP	JIP-AOSRT	2013	The objective is to conduct a workshop that identifies as possible step change in recovering spilled oil in ice-infested waters (completed March 2012).
	"Oil in Ice JIP" Testing of Oil Skimmer via Field Experiments in the Barents Sea, May 2009		SINTEF & CounterSpil Research	JIP SINTIF Oil In Ice	2010	The main objective of the field testing was to test and verify the Ro-Clean Desmi and Framo skimmers in the field under more realistic conditions compared to what can be accomplished in the basin testing. This includes lower oil, temperature fluctuations, wind, and a more dynamic ice field. This report covers the of mechanical recovery equipment performed during the experimental field trial in the Barents Sea from May 7–23, 2009. http://www.sintef.no/project/JIP_Oil_In_Ice/Dokumenter/publications/JIP-rep-no-21-Field%20report%202009-mechanicalfinal_2010.pdf

(Continued)

Research Initiatives	Description	Facility	PI	Funding	Year	Objective
	Testing and Verification of Oil Skimmers During the Field Experiment in the Barents Sea , May 2008		SINTEF	JIP SINTIF Oil In Ice	2010	The work reported herein was part of an experimental field trial in the Barents Sea from 18 to 30 May 2008. The field trial was performed in the eastern part of the Barents Sea east of the island of Hopen. Three oil recovery skimmers were tested during this field trial over a period of two days. http://www.sintef.no/project/JIP_Oil_In_Ice/Dokumenter/publications/JIP-rep-no-9-Field%20report%202008-mechanicalfinal_2010.pdf
	Testing of Ro-Clean Desmi Polar Bear Skimmer in SINTEF Ice Basin.	SINTEF	SINTEF, SL Ross, & DF Dickins Asscoiates	JIP SINTIF Oil In Ice	2010	This report summarizes the findings of a test program conducted in September 2008 in the SINTEF ice basin of the Polar Bear skimmer. The skimmer is a mechanical oil recovery device that was designed on the basis of 2007 tests in the

Research Initiatives	Description	Facility	PI	Funding	Year	Objective
	Task 3.2 Testing of New Concepts and Units					SINTEF ice basin of Ro-Clean Desmi's Helix Skimming adapter and Ice Skimmer as well as subsequent field evaluations of the Helix in the Barents Sea in May 2008. http://www.sintef.no /project/JIP_Oil_In_Ice/Dokumenter/publica tions/JIP-rep-no-10-Desmi%20Polar%20Bear-Basin%20test2008-final_2010.pdf
	Testing of Lamor GT 185 Skimmer and LRB 150 Skimmer in SINTEF Ice Basin. Task 3.1.Testing of Existing Concepts. A Technical Report	SINTEF	SINTEF	JIP SINTIF Oil In Ice	2010	The main objective of this project was to document the capability and potential application of commercially available skimmers for recovering oil in ice. Based on this documentation, suggestions should be possible for defining and improving the operational spill response window in ice and cold conditions. http://www. sintef. no/project/JIP_Oil_In_Ice/Dokumenter/publica tions/JIP-rep-no-8-Report-Lamor-task%203.1-final_2010.pdf

(Continued)

Research Initiatives	Description	Facility	PI	Funding	Year	Objective
	Testing of Ro-Clean Desmi Ice Skimmer and Helix Skimmer in SINTEF Ice Basin. Task3.1: Testing of Existing Concepts	SINTEF	SINTEF	JIP SINTIF Oil In Ice	20010	The main objective of this project was to document the capability and potential application of commercially available skimmers for recovering oil in ice. Based on this documentation, suggestions should be possible for defining and improving the operational spill response window in ice and cold conditions. http://www.sintef.no/project/JIP_Oil_In_Ice/Dokumenter/publications/JIP-rep-no-7-Report-Desmi-task%203.1-final_2010.pdf
	Fate of Emulsion Breakers Used for Decanting	Ohmsett	SL Ross	MMS	2008	Determine the partitioning of different chemical emulsions breakers between oil and water phases when they are used to enhance decanting of recovered water from offshore. See 395 skimming operations.

Research Initiatives	Description	Facility	PI	Funding	Year	Objective
	Testing and Evaluation of Sorbents		SAIC-Canada	MMS & EC	2007	Performance of sorbent materials used in the remediation of oil spills. http://www.etc-cte.ec.gc.ca/databases/Sorbent/Default.aspx.
	Oil Recovery with Novel Skimmer Surfaces under Cold Climate Conditions	CRREL	U of C-SB / Ohmsett Staff	MMS	2007	The objective of this project was to perform a comprehensive analysis of the adhesion processes between oil or ice-in-oil mixtures and various surface patterns and materials that are being used or proposed for use in oil skimmers, under cold climate conditions.
	Tailored Polymeric Materials for Oil Spill Recovery in Marine Environments		U of C-SB	MMS	2006	This project substantially increased the efficiency of mechanical oil spill recovery equipment by replacing traditional recovery unit materials with polymetric materials that have the highest affinity for oil and are specifically tailored to collect oil from water surfaces.

(Continued)

Research Initiatives	Description	Facility	PI	Funding	Year	Objective
	Optimization of Oleophilic Skimmer Recovery	OHMSETT	U of C-SB	MMS	2006	Existing oleophilic skimmers collect oil at a relatively slow rate, in large part because the selection of materials used for the adhesion surfaces of the oil recovery units has not been based on their adhesion properties, but rather on historical practice or availability. See TAR 511.
	Evaluation of Skimmers for Offshore and Ice-Infested Waters		EC-EED	MMS & EC	1993	Innovative skimming technology for oil spill cleanup offshore and in icy waters. Current skimmer technology for recovering oil in broken ice conditions was practically nonexistent.
	Field Evaluation of Oil Spill Chemical Additives		EC-C&PA	MMS &EC	1990	Evaluated two new oil spill chemical additives during a series of intentional oil spills off the Canadian coast during September 1987 to mechanical recovery.

Research Initiatives	Description	Facility	PI	Funding	Year	Objective
	Surface Oil Spill Containment and Cleanup		Veritas Technical Services, Inc.	MMS	1987	Engineering concept analysis of the effectiveness of a large, self-contained oil spill collection ship capable of deploying large skimming booms in the proximity of a blowing oil well.
	Subsea Collection of Blowing Oil and Gas		Brown and Root	MMS	1985	The prospects are attractive for using large, self-contained collection ships which can deploy subsea collectors.
Deep Well	Project "Deep Spill"		SINTEF	MMS and 23 different oil companies to conduct this research	2005	Simulate a blowout or pipeline rupture in deep water and obtain data to verify the predictions of a deep water blowout model.
	Oil Spill Containment, Remote Sensing, and Tracking from Deep Water Blowouts Status of Existing and Emerging Technologies		PCCI Marinenand Environ-mental Engineering	MMS	1999	Assessment of existing or developing technologies that could be used to sense, track, contain, and recover oil released by deep water blowouts or pipeline ruptures.

(Continued)

Research Initiatives	Description	Facility	PI	Funding	Year	Objective
	Project Cancelled: In Situ Burning on Mud Flats		SINTEF	MMS	2010	Investigate the feasibility of using in situ burning to remove crude oil from mud flats in an environmentally and ecologically favorable manner.
Beach	In Situ Clean up of Oiled Shorelines; Svalvard Shoreline Project		EC	MMS, EC, Texas General Land Office, Norwegian Pollution Control Authority, Exxon Imperial Oil Resources Ltd., UK Marine Pollution Control Unit, Swedish Rescue Service, and the CCG.	1999	Investigated the effectiveness of mainstream in situ shoreline cleanup techniques (tilling, surf-washing), as well as some of the scientific aspects of oil behavior and oil removal from shorelines by what was commonly referred to as the clay-oil flocculation process.

Research Initiatives	Description	Facility	PI	Funding	Year	Objective
	Development and Evaluation of Shoreline Cleanup Techniques		EC-W-TDB	MMS & EC	1994	Development and Evaluation of Shoreline Cleanup Techniques
Micro-Biological Remedi-ation	Oil Distribution and Bioavailability Field Experiment–FEX 2009		SINTEF	JIP SINTIF Oil In Ice	2010	GPS trackers, large-volumes water samplers (KISPs), in situ oil-in-water monitoring systems and passive absorption devices (SPMDs) were all installed under ice floes in and around the oil slick to enable a detailed monitoring of oil-inwater dynamics and ice interactions throughout the six-day experiment. Meteorological and oceanographic data were recorded for the monitoring of wind speed and direction, air temperature, currents, and ice movement. http://www.sintef.no/project/JIP_Oil_In_Ice/Dokumenter/publications/JIP-rep-no-33-FEX2009%20Oil%20distribution%20and%20bioavailability.pdf

(Continued)

Research Initiatives	Description	Facility	PI	Funding	Year	Objective
	Oil Biodegradation in Arctic ice		SINTEF		2005	Biodegradation is one of the most important processes for the removal of petroleum hydrocarbons from marine environments. Also, in Arctic environments, biodegradation is important, and during summer conditions degradation follows a pattern indistinguishable from that exhibited under temperate climate conditions. However, the necessary knowledge about hydrocarbon biodegradation of oil frozen in Arctic ice is lacking. The ice contains a variety of cold-adapted microorganisms, and metabolic activities have been measured at ice temperatures well below –10ÀC.

Research Initiatives	Description	Facility	PI	Funding	Year	Objective
Testing Standards						
	Laboratory Testing to Determine Dispersion Predictability of the Baffled Flask Test (BFT) and Swirling Flask Test (SWT)		US EPA / University of Cincinnati	MMS	2006	Conduct standardized laboratory dispersant effectiveness testing to determine if either the Baffled Flask Test (BFT) or the Swirling Flask Test (SWT) is able to predict chemical dispersion effectiveness of oil spills.
	Development of a Draft ASTM Standard on Characterizing Oils for In Situ Burning		SL Ross	MMS	2005	Developed a draft ASTM guidelines for in situ burning from lab analysis of the oil to determine how easily it can be burned.
	Procedures for Reporting Tests of Oil Spill Containment Booms and Skimmers		Environ-mental Consultant, Inc.	MMS	2005	Reviewed existing test procedures for oil containment booms and skimmers and developed standardized test reporting procedures that will eliminate the problems of incomplete test documentation.

(Continued)

Research Initiatives	Description	Facility	PI	Funding	Year	Objective
	Development of a Standard Method for Measuring the Buoyancy-to-Weight Ratio for Oil Spill Containment Boom	OHMSETT	SL Ross	MMS	2005	The objective of the work was to develop a standard method for measuring the buoyancy-to-weight ratio of oil spill containment booms.
	Development of a Method to Produce Large Quantities of Realistic Water-In- Oil Emulsions for use in Evaluating Oil Spill Response Equipment and Methods		SL Ross	MMS	2005	Developed a methodology and equipment to enable researchers to produce large quantities of realistic water-in-oil emulsions.

Research Initiatives	Description	Facility	PI	Funding	Year	Objective
	Preparation a Test Protocol for Offshore Oil Skimmers and Containment Booms		Chapman	MMS & EC	1992	Development of a nonpolluting, cost-effective standard test procedure for evaluating the performance of open ocean containment booms and oil skimmers is needed. (1) Test Protocol for the Evaluation of Oil-Spill Containment Booms (February 1992) and (2) Suggested Test Protocol for the Evaluation of Oil Spill Skimmers for the OCS (February 1992).
	Oil Spill Response Equipment Performance Verification	OHMSETT	Frank Weston and Associates	EPA	1990	Develop and verify an innovative test procedure for measuring the performance of offshore oil spill response equipment at sea.
	Open Ocean Boom Test		EC-C&PA	MMS & EC	1990	Develop a nonpolluting and cost-effective testing procedure for recovery equipment.

(Continued)

Research Initiatives	Description	Facility	PI	Funding	Year	Objective
Recent Field Demonstrations (post 2000)						
	Countermeasure Verification Through Controlled Field Releases and Exercises		OGP	JIP-AOSRT	2012	To create opportunities for the JIP technical working groups to test, evaluate, and verify selected technologies and to conduct related research (for example, environmental effects) in a field setting.
	Field experiment in the St. Lawrence River estuary		COOGER & CCG		2008	A small spill was carried out in pack ice from a Coast Guard icebreaker and immediately treated by spraying OMA and agitating the oil dispersant mixture.
	Full scale field experiment 2009 Cruise report		SINTEF	JIP SINTIF Oil In Ice	2009	This report gives an overview of the activities and main findings during the full-scale field experiment carried out in the period May 9–25, 2009, in position N77.30 - E30.90 east of Hopen in the Barents Sea.

Research Initiatives	Description	Facility	PI	Funding	Year	Objective
						http://www.sintef.no/project/JIP_Oil_In_Ice/Dokumenter/publications/JIP-rep-no-25-Field-report-FEX%202009-final.pdf
	Field experiments in the Norwegian Barents Sea - East of Svalbard		SINTEF Oil in Ice JIP	AGIP KCO, Chevron, Conoco Phillips,Shell, Statoil & Total	2008/09	Series of experimental spills carried out from research vessel and icebreaker. Response techniques included herding agents, burning in situ and in fireproof booms, dispersants, mechanical systems, and remote sensing
	Planning Support for an Experimental Oil Spill in Pack Ice. Project was canceled a due to CGC rescue mission		SL Ross & Dickins	MMS	2008	Experimental oil spill was to be conducted by the Department of Fisheries and Oceans (DFO) Canada offshore eastern Canada in March 2007. The Canadian Coast Guard (CCG) was committed to supplying an ice breaker and helicopter as support vessels for one week as services in kind to the project.

(Continued)

Research Initiatives	Description	Facility	PI	Funding	Year	Objective
	Full-Scale Field Experiment 2008, Summary of Results		SINTEF	JIP SINTIF Oil In Ice	2008	Report is an overview of the activities and main findings during the full-scale field experiment carried out in the period May 18–28, 2008, in position N77.30–E30.90, east of Hopen in the Barents Sea. http://www.sintef.no/project/JIP_Oil_In_Ice/Dokumenter/publications/JIP-rep-no-18-FEX-2008-field-results-summary.pdf
	FEX 2008. Activity Report		SINTEF	JIP SINTIF Oil In Ice	2008	Report is an overview of the 2008 JIP oil in ice full-scale filed experiment http://www.sintef.no/project/JIP_Oil_In_Ice/Dokumenter/publications/JIP-rep-no-17-FEX-2008-field-activity-report.pdf

Research Initiatives	Description	Facility	PI	Funding	Year	Objective
	Svalbard, Norway Experimental Oil Spill To Study Spill Detection and Oil Behavior in Ice		DF Dickins Asscoiates, SINTEF, BSU, The University Center at Svalbard	MMS, ADEC, ACS, Conoco Phillips, ExxonMobil, Shell Technology Norway, and Statoil ASA	2006	This project involved spilling 3,400 liters of crude oil within a skirt inserted through fast ice at Svea on Svalbard. Acoustic and GPR systems (surface and airborne) were tested for there abilities to detect oil under ice. Oil behavior was studied to look at migration rates, dissolution into the water column and weathering. The oil was removed successfully by burning.
	Oil-in-Ice on Svalbard		SINTEF	Norwegian Research Council, Statoil, and Hydro	2005	The purpose of this three-year project was to investigate weathering processes in marine oil spills under Arctic conditions. The focus was on weathering of the bulk oil phase (evaporation, w/o emulsification, and dispersion), dissolution of water-soluble components, photo-oxidation and biodegradation processes of oil in ice.

(Continued)

Research Initiatives	Description	Facility	PI	Funding	Year	Objective
Miscellaneous						
	Characteristics, Behavior and Response Effectiveness of Spilled Dielectric Insulating Oil in the Marine Environment		Louisiana State University		2011	of wind turbine generators connected to a centralized electrical service platform (ESP). The ESP could contain approximately 40,000 gallons of dielectric insulating oil and approximately 2,000 gallons of assorted oil-based fluids (diesel fuel, lubricating oils, etc.) stored on site for facility maintenance.
	Ohmsett Biofuel Feasibility Study		SL Ross	BSEE	2011	BioFuels (bio-diesels, and gasoline/ethanol blends) in the worldwide marketplace. The different physical and chemical properties of these products in comparison to the petroleum products they are replacing leads to the issue of response countermeasure actions in the event of a BioFuel spill.

Research Initiatives	Description	Facility	PI	Funding	Year	Objective
						In order to pursue testing and implementation of BioFuel response techniques at Ohmsett, the facility's operating systems and protocols for traditional oil spill response must be reviewed, compared, and possibly modified, to adapt to the potential requirements for BioFuel response techniques testing.
	The Industry/University Cooperative Research Center for Multiphase Transport Phenomena		Michigan State University and the University of Tulsa	BSEE	2010	Multiphase Transport Phenomena has focused on two technologies important to DOI/MMS: (1) cross flow filtration hydrocyclones for oil/water separation (onshore, offshore, subsea, and downhole) and (2) the use of computational methods to address flow assurance issues associated with oil and gas production.
	Synthetic Based Drilling Fluid Droplet Survey		Southwest Research Institute	MMS	2007	Synthetic-based drilling muds (SBMs) have been developed to provide the desirable operating qualities of oil-based muds with the

(Continued)

Research Initiatives	Description	Facility	PI	Funding	Year	Objective
						lower toxicity of water-based muds. The objective of this project was to evaluate the fall velocity distributions that would result from an accidental release of SBMs in offshore waters.
	Identification of Window of Opportunity for Chemical Dispersants on Gulf of Mexico Crude Oils		SL Ross	MMS	2007	Develop best-fit correlations between readily available fresh oil properties and the window of opportunity for successful chemical dispersant use on Gulf of Mexico crude oils.
	Process for the Removal of Spent Oil Spill Dispersants from Test Water at Ohmsett	Ohmsett	SAIC Canada	MMS	2004	The project identified potential methods for the removal of dissolved dispersants from Ohmsett tank water using membrane filtration technology. The goal was to lower the dispersant concentration in the tank water to undetectable levels after dispersant effectiveness testing.

Research Initiatives	Description	Facility	PI	Funding	Year	Objective
	Wave Field Characterization at the Ohmsett Wave Test Basin	Ohmsett	U of Wa	MMS	2004	The project was designed to find out the characterizations of waves in the Ohmsett Wave Basin facility.
	Techniques to Remove Dissolved Dispersant from Ohmsett Basin Water	Ohmsett	SL Ross	MMS	2003	Researched techniques to remove dissolved dispersant from Ohmsett water after the tank has been used for a series of chemical dispersant effectiveness experiments.
	Development of an Ohmsett Activity Summary Report	Ohmsett	Marine Research Associates LLC	MMS	1998	Develop an effective means of summarizing and disseminating the testing capabilities, protocols, and test results for the MMS Ohmsett.
	Identification of Substitute Test Facilities for Ohmsett		EC-EED	MMS	1990	EPA suspended operation of the Ohmsett facility, and alternative facilities and procedures are necessary to minimize the delay in testing innovative oil spill response technology and procedures.
	Oil and Gas Transportation Russia- Europe		SINTEF			Research and development project supported by the European Union.

(Continued)

Research Initiatives	Description	Facility	PI	Funding	Year	Objective
						The overall objective of ARCOP is to form an operational platform for the development of oil and gas transportation from the Russian Arctic (the Pechora and Kara Sea) to Europe. http://www.sintef.no/home/Materials-and-Chemistry/Marine-Environmental-Technology/Projects-and-News/Oil-and-gas-transport-Russia---Europe-/

Acronyms Used in "An Inventory of Research Projects on Oil Spills in Arctic Waters"

Acronym	Summary	Web Site
ACS	Alaska Clean Seas: ACS protects the environment by providing effective response services to the Alaska North Slope Crude Oil Producers and the first 167 miles of the Trans-Alaska Pipeline System	http://www.alaskacleanseas.org/
AKEC-SPAR	Alaska Department of Environmental Conservation Spill Prevention and Response (SPAR) prevents spills of oil and hazardous substances, prepares for when a spill occurs and responds rapidly to protect human health and the environment.	http://dec.alaska.gov/spar/index.htm
BOEM	Bureau of Ocean Energy Management: Protecting the environment while ensuring the safe development of the nation's offshore energy and marine mineral resources is a critical part of BOEM's mission.	http://www.boem.gov/
BOEMRE	Bureau of Ocean Energy Management, Regulation and Enforcement	http://www.boemre.gov/
BSU	Boise State University, John Bradford	http://cgiss.boisestate.edu/~johnb/Pubs.html
BSEE	Bureau of Safety and Environmental Enforcement works to promote safety, protect the environment, and conserve resources offshore through vigorous regulatory oversight and enforcement.	http://www.bsee.gov/
CCG	Canadian Coast Guard	http://www.ccg-gcc.gc.ca/eng/CCG/Home?is mobile=2
CISPRI	Cook Inlet Spill Prevention & Response Inc. (CISPRI) Team provides our Member companies with the most professional, cost-effective contingency planning and spill removal possible.	http://www.cispri.org/

(Continued)

Acronym	Summary	Web Site
COOGER	Centre for Offshore Oil, Gas and Energy Research (COOGER). Fisheries and Oceans Canada established COOGER in 2002 to coordinate the department's nation-wide research into the environmental and oceanographic impacts of offshore petroleum exploration, production, and transportation.	http://www.dfo-mpo.gc.ca/science/coe-cde/cooger-crpgee/indexeng.htm
CRREL	US Army Cold Regions Research and Engineering Laboratory	http://www.crrel.usace.army.mil/ http://www.crrel.usace.army.mil/innovations/oil_spill_research/
Dickins	DF Dickins and Associates, LLC	http://www.dfdickins.com/
DFO	Department of Fisheries and Oceans Canada (DFO) and its Special Operating Agency, the Canadian Coast Guard, deliver programs and services that support sustainable use and development of Canada's waterways and aquatic resources.	http://www.dfo-mpo.gc.ca/index-eng.htm
EC	Environment Canada is a diverse organization where our programs, services, and people lead the way in implementing the Government of Canada's environmental agenda. We collaborate with our partners at home and abroad, to realize concrete progress on initiatives that will protect the health of our people and our planet.	http://www.ec.gc.ca/default.asp?lang=En&n=BD3CE17D-1
EC-C&PA	Conservation and Protection Agency, Environment Canada	
EC-EED	Emergencies Engineering Division, Environment Canada	http://www.ec.gc.ca/ee-ue/
EC-ESD	Emergencies Science Division, Environment Canada	http://www.etc-cte.ec.gc.ca/organization/estd e.html

Acronym	Summary	Web Site
EC-W-TDB	Western Office, Technology Development Branch, Environment Canada	
EPA	EPA seeks to prevent, prepare for, and respond to oil spills that occur in and around inland waters of the United States. EPA is the lead federal response agency for oil spills occurring in inland waters, and the US Coast Guard is the lead response agency for spills in coastal waters and deepwater ports.	http://www.epa.gov/oilspill/
JIP-AOSRT	Arctic Oil Spill Response Technology (AOSRT)–Joint Industry Programme. An international research program to further enhance industry knowledge and capabilities in the area of Arctic oil spill response.	http://www.arcticresponsetechnology.org/
JIP-API	American Petroleum Institute and the Joint Industry Programme on Oil Spill Recovery in Ice	
JIP-SINTEF		http://www.sintef.no/Projectweb/JIP-Oil-In-Ice/
MMS	Mineral Management Services evolved to BOEMRE and subdivided to BOEM and BSEE	
Ohmsett	The National Oil Spill Response Research & Renewable Energy Test Facility provides independent and objective performance testing of full-scale oil spill response equipment and marine renewable energy systems (wave energy conversion devices), and improving technologies through research and development.	http://www.ohmsett.com/
OSRI	Prince William Sound Oil Spill Recovery Institute	http://www.pws-osri.org/

(Continued)

Acronym	Summary	Web Site
PERF	Petroleum Environmental Research Forum (PERF)was established in 1986 to stimulate cooperative research & development of technology for environmental pollution control & waste treatment for the petroleum Industry.	http://www.perf.org/
SL Ross	SL Ross Environmental Research Ltd.	http://www.slross.com/
SAIC Canada		http://www.companylisting.ca/SAIC Canada1/default.aspx
SAMS	Scottish Association for Marine Science	http://www.smi.ac.uk/
SINTEF	SINTEF is the largest independent research organization in Scandinavia. Its goal is to create value through knowledge generation, research, and innovation, and develop technological solutions that are brought into practical use.	http://www.sintef.no/home/ http://www.sintef.no/home/Materials-and-Chemistry/MarineEnvironmental-Technology/Fate-and-effects/
Statoil		http://www.statoil.com/en/Environment Society/Sustainability/2007/Envi ronment/Environment/Research/Pages/ OilInNorth.aspx
TAS	Thales Alenia Space	http://www.thalesgroup.com/Markets/S pace/Related_Activities/Thales_ Alenia Space/
TC	Transport Canada	http://www.tc.gc.ca/
UNH-CRRC	Costal Response Research Center at University of New Hampshire	http://www.crrc.unh.edu/

Acronym	Summary	Web Site
USCG	US Coast Guard	http://www.uscg.mil/default.asp
WHOI	Woods Hole Oceanographic Institution	http://www.whoi.edu/

In: Oil Spills in Arctic Waters
Editor: Nora Linwood
ISBN: 978-1-63321-580-1

Chapter 3

TABLE B: ALASKA DEPARTMENT OF ENVIRONMENTAL CONSERVATION, DIVISION OF SPILL PREVENTION AND RESPONSE, RESEARCH AND DEVELOPMENT PROJECTS (2000–2010)*

United States Arctic Research Commission and US Army Corps of Engineers Cold Regions Research and Engineering Laboratory

YEAR	PROJECT TITLE	PROJECT DESCRIPTION
1997–2005	Environmental Sensitivity Index (ESI) Mapping	Series of seasonal maps intended to provide a regional overview of the environmentally sensitive resources that would receive priority protection during oil spill planning and response. The resources depicted on the maps are selected on the basis of their high sensitivity and/or vulnerability to spilled oil or their special management status. The following is a summary of the ESI mapping status by subarea: Aleutians (2002) Bristol Bay (2004) Cook Inlet (2002) Kodiak (1997) North Slope (2005) Northwest Arctic (2002)

* This is an edited, reformatted and augmented version of an online table that accompanies the white paper found in Chapter 1, issued November 2012.

Table B. (Continued)

YEAR	PROJECT TITLE	PROJECT DESCRIPTION
		Prince William Sound (2000) Southeast Alaska (2001) Western Alaska (2003) More information available at: http://response.restoration.noaa.gov/book_shelf/827_ERD_ESI.pdf
2000	International Oil & Ice Workshop, 2000	Workshop on oil spill preparedness and response for cold climates. Provided an opportunity to congregate experts on oil fate and behavior, Arctic oil spill response, ice environments, and Arctic oilfield development in order to present the state of knowledge in a combined classroom and field setting (Anchorage and Prudhoe Bay).
2000	MORICE Phase 5	Mechanical Oil Recovery in Ice Infested Waters (MORICE) was a multiphase Joint Industry Program (JIP) to develop technologies for more effective recovery of oil spills in ice-infested waters. Phase 5, evaluated four different internal oil recovery units for the lifting grated belt at a test tank in Hamburg, Germany, in May 2000. Upon completion of these tests, the prototype skimmer was shipped back to Prudhoe Bay, AK, where it received further equipment modifications, and more powerful hydraulics and a larger power pack were installed. Selected skimmer manufacturers were invited to participate and provide an internal recovery system for evaluation. The complete prototype skimmer was evaluated for ice processing on the Alaskan Beaufort Sea during October 2000.
2000	Oil Detection in Ice	Pangaea Geochemical Technologies demonstrated the potential use of gas-sieve gas chromatograph instrumentation capable of detecting molecular levels of hydrocarbons in ice core samples.
2000	Barge Ice Deflection System	Design and construct a Barge Ice Deflection System (BIDS). The concept, design, fabrication, and testing of BIDS was a direct result of lessons learned from fall 1999 drilling on the North Slope. Spill responders recognized the need to deflect ice away from barge skimming systems in order to maintain a reasonable recovery capacity.
2000	Temporary Pipeline Concept	As a result of lessons learned during a barge offloading demonstration on the North Slope, a study was implemented to examine the feasibility, design, process, and rough order-of-magnitude costs for a system capable of handling recovered oil spill fluids consistent with the response planning standard (RPS) requirements from the Northstar C-Plan.
2000	Viscous Oil Pumping Demonstration	Previous tests indicated the injection of water through an annular injection flange significantly reduced the discharge hose pressures while pumping viscous fluids. This test investigated the potential to adapt this technology to a cold climate through injection of alternative lubricating fluids.
2000	Emulsion Breaker Study, 1	Determined the effect of small dosages of emulsion-breaking chemical on several North Slope crude oil emulsions at or near freezing temperatures.
2000	Crude Oil Evaporation Study	Study to provide an educated analysis of the potential evaporative losses from an oil and gas surface well blowout from the Norhstar and Pt. McIntyre fields on the Alaskan North Slope.

YEAR	PROJECT TITLE	PROJECT DESCRIPTION
2000–2002	GNOME Oil Trajectory Model for the Alaskan North Slope	In coordination with NOAA, this project worked to develop a regional oil spill trajectory model appropriate for use in the nearshore marine environment of the Alaskan Beaufort Sea. GNOME, the General NOAA Oil Modeling Environment, is a general-purpose oil spill trajectory model used by NOAA/HAZMAT.
2001 and ongoing	Geographic Response Strategies (GRS)	GRS are oil spill response plans tailored to protect a specific sensitive area from oil impacts following a spill. These response plans are map-based strategies that can save time during the critical first few hours of an oil spill response. They show responders where sensitive areas are located and where to place oil spill protection resources. The following is a summary of the GRS initiatives thus far by subarea: Aleutians (2004 and ongoing) Bristol Bay (project to be initiated over next 1–2 years) Cook Inlet (2001 and ongoing) Kodiak (2001 and ongoing) North Slope (project to be initiated over next 1–2 years) Northwest Arctic (project to be initiated over next 1–2 years) Prince William Sound (2002 and ongoing) Southeast Alaska (2002 and ongoing) Western Alaska (project to be initiated over next 1–2 years) More information available at: http://dec.alaska.gov/spar/perp/grs/home.htm
2001	Ice Access Guidelines for Spill Responders	Sandwell Engineering Inc. conducted research and developed a manual focused on oil spill response efforts to support both on-ice and under-ice cleanup activities on the North Slope. The manual provides equipment operators, site managers, and project managers the information necessary to understand the factors and requirements to maintain safety during on-ice operations.
2001	MORICE Phase 6	The prototype skimmer (from Phase 5) was considered ready for oil in ice testing. The unit was shipped to Svalbard, Norway, in May 2001 for testing, however, due to complications, the unit was not field-tested. The MORICE project concluded after Phase 6A which successfully tested and evaluated the prototype skimmer at Ohmsett in January 2002. The unit was tested in cold water and broken ice conditions with oil and two different internal recovery systems.
2001	National Institute of Standards & Technology (NIST) Well Blowout Dispersion Modeling	Contract with NIST to create a 3-D model of a drilling related blowout. The model includes the presence of a rig structure and can account for variations in flow rate, oil properties, gas-to-oil ratios, discharge pressures, wind conditions, and impingement effects of drill rig structure on the oil and gas plume.
2002	Alaska Incident Management System (AIMS) Guide	The AIMS Guide was developed for use by public and private agencies to fully coordinate response efforts during a significant oil or hazardous materials release. While the roles of the government organizations may vary from directing the response, augmenting the response, or providing regulatory oversight, the intent of this document is to foster a common understanding of the roles and responsibilities of all responding agencies to ensure a safe, effective response. More information is available at: http://dec.alaska.gov/spar/perp/docs/AIMS_Guide-

Table B. (Continued)

YEAR	PROJECT TITLE	PROJECT DESCRIPTION
		Complete(Nov02).pdf
2003	NIST, Well Blowout Droplet Size Study	Preliminary study to investigate the drop size distributions created from various pipe diameters under a range of flow rates and GORs.
2003	Viscous Oil Pumping Studies, USCG	Joint Viscous Oil Pumping System (JVOPS) Workshop, headed by representatives of the Canadian and US Coast Guards was held in Houma, LA, Dec 1–15, 2003. The purpose of the workshop was to improve oil spill response systems for heavy viscous oils. The workshop targeted testing in the critical 200,000–500,000+ cSt range and pumping distances up to 450 meters. A range of pumping systems was also tested.
2003	Viscous Oil Pumping Studies, Mini Barge	In a test, a vacuum truck pumped a 249 bbl mini barge of its cargo of emulsified Northstar crude oil at 32°F in 46 minutes. The pumping confirmed long-standing planning assumptions regarding transfer rates of emulsified oil at ambient sea temperatures stored by on-water spill response systems.
2003	Emulsion Breaker Study, 2	Test to examine the fate of emulsion breaker chemicals that may be used to help decant free water from oil storage barges.
2004	Testing Portable Pumps with Viscous Oil	Demonstrated pumping capability of ACS equipment for recovering oil spilled to tundra in cold conditions. The key elements of this test were to measure the rate at which weathered oil could be recovered with small, portable pumps and to determine the limits of pumping related to pour points of the oil.
2004 – 2006	NIST Well Blowout Model Development	Continue National Institute of Standards and Technology's (NIST) well blowout model development. Develop Alaska-based capability to apply and run NIST's oil dispersion model for cases of rig and no rig and fire and no fire. Compare the results of a no-fire-no-rig case to SL Ross's model results. Develop capability to model the affect of a plume deflector.
2004	TAP II Modeling, North Slope	Trajectory Analysis Planner (TAP II) Modeling. GNOME model location files for Alaskan Beaufort Sea provided opportunities to expand spill trajectory modeling to statistically-based projections of potential spill impact areas. The graphical output from TAP II includes threat zone analysis, shoreline impact analysis, site oiling analysis and resource impact analysis.
2004 – 2005	Testing Emulsion Breakers to Improve Pumping of Viscous Emulsions	Quantify the effectiveness of demulsifier addition in reducing emulsion viscosity, and hence friction-induced backpressures, when pumping emulsions in hoses.
2004 – 2006	North Slope Coastal Ocean Dynamic Applications Radar (CODAR)	Support for a project led by the University of Alaska Fairbanks to develop a land based radar system for measuring the speed and direction of the Beaufort Sea's surface currents.
2004 – 2006	Ground Penetrating Radar	Studies of GPR for the detection of oil in and under sea ice: November 2004 experiments at the U.S. Army Cold Regions Research

YEAR	PROJECT TITLE	PROJECT DESCRIPTION
	(GPR) for Detecting Oil in and Under Ice	and Engineering Laboratory (CRREL) in Hanover, NH; April 2005 Prudhoe Bay, AK, field experiments (without oil at temperatures -20 degrees F); March 2006 Svalbard, Norway, field experiments with oil under ice.
2005	North Slope Nearshore & Offshore Breakup Study	Provides an analysis of ice conditions and the dates associated with seasonal transitions in the nearshore and offshore Beaufort Sea environments including the lagoon areas inside the barrier islands affected by major river overfloods. More information available at: http://dec.alaska.gov/spar/ipp/nscharter.htm
2005	Viscous Oil Pumping Studies	Continuation of previous work. Tested a range of North Slope pumps in North Slope liquids to determine their capability. Tested a range of additional annular injection unit types. Measured viscosity and pump rates resulting from treatments of a range of oils with a range of emulsion breakers. Developed pump well and lids accommodating hard pipe sections and cargo level gauging for mini barges. Developed ready-to-go prototype annular injection packages for skimmer pump, mini barge pump and vacuum truck pump.
2005	ADEC Tundra Treatment Manual	Updated the manual with new tactics. The purpose of this manual is to provide a menu of tactics which can be used to treat and monitor tundra impacted by spills of crude oil, petroleum products, seawater, and other substances after initial response efforts have eliminated the threat of large-scale spill migration. More information is available at: http://dec.alaska.gov/spar/perp/r_d/ttman/tt_man.htm
2005–2006	Spill Related Analysis of North Slope Crude Oils	Oil characteristics affecting recovery and transfer planning were determined by laboratory tests. The viscosity, emulsification tendency, specific gravity, and evaporation rates of crude oils from 11 North Slope sources were measured.
2005–2009	SINTEF Oil in Ice Joint Industry Program (JIP) Support	Provided support through Alaska Clean Seas. The objective of the JIP Oil-in-Ice project led by the Norwegian research group SINTEF was to "further develop tools and technologies for environmental beneficial oil spill response strategies for ice-infested waters." The project culminated with a large-scale field trial with actual releases of oil in ice-infested waters (FEX 2009). The FEX 2009 was carried out in the period May 9–25, 2009, east of Hopen, Norway, in the Barents Sea.
2006	Alaska Commercial Fisheries - Water Quality Sampling Methods and Procedures Manual	This manual presents scientific water-quality sampling methods to assess potential contamination of commercial fishery resources and gear in Alaska waters during an oil spill. It is intended for oil spill response personnel and fishery managers as they assess and manage the risk to commercial fisheries during spill events. The manual presents both a general discussion of factors for consideration in designing a sampling program after a spill event, and a description of specific methods and procedures. More information available at: http://dec.alaska.gov/spar/perp/wq/wq_manual.htm
2006	Spill Tactics for	This project led to the development of a statewide spill response tactics

Table B. (Continued)

YEAR	PROJECT TITLE	PROJECT DESCRIPTION
	Alaska Responders (STAR) Manual	manual for use by the spill response community, including federal, state, local, industry, and spill cooperatives throughout Alaska. The final product is available for general use by the spill response community in Alaska, and also serves as a means for the oil industry to meet contingency planning requirements. More information available at: http://dec.alaska.gov/spar/perp/star/index.htm
2006	De-Ice Rope Mop Skimmer Studies	Methods of de-icing rope mop skimmers in cold temperatures were examined. The research included a literature review of the state-of-the-art in de-icing technology, and a series of small-scale tests at the SL Ross laboratory to evaluate candidate technologies.
2006	Pour Point Depressants Studies	Measure the affect of pour point depressors on the viscosity and "pump-ability" of cold, viscous recovered and stored oil and oil emulsions. Pour point depressors are introduced in ppm. They are typically used in some oil production processes, but had not yet been examined for spilled oil transfer.
2006 and ongoing	Potential Places of Refuge (PPOR)	PPOR are pre-identified sites that may aid decision-makers in responding to vessels in distress. These plans are tailored to protect sensitive areas from impacts from possible spills and are map-based to save time during the critical first few hours of a vessel response. The following is a summary of the PPOR initiatives thus far by subarea: • Aleutians (completed in 2007) • Bristol Bay (project to be initiated over next 1–2 years) • Cook Inlet (completed in 2007) • Kodiak (completed in 2006) • North Slope (project to be initiated over next 1–2 years) • Northwest Arctic (project to be initiated over next 1–2 years) • Prince William Sound (completed in 2008) • Southeast Alaska (2008, to be completed in 2010) • Western Alaska (project to be initiated over next 1–2 years) More information is available at: http://dec.alaska.gov/spar/perp/ppor/home.htm
2007	Vessel Emergency Towing System (ETS)	Develop emergency towing capabilities for disabled vessels using locally available tugboats in conjunction with ETS equipment that may be stationed in strategic locations in Alaska or air deployable via US Coast Guard or other assets. • Unalaska (2007; two systems in place) • Kodiak (2009) • Two additional systems to be purchased in 2010–2011 (Southeast Alaska; Anchorage air deployable system) More information is available at: http://dec.alaska.gov/spar/perp/aiets/home.htm
2007	International Oil & Ice Workshop,	Workshop on advancing spill response in cold water and ice. Provided an opportunity to bring together an international audience with this

YEAR	PROJECT TITLE	PROJECT DESCRIPTION
	2007	common interest. The two-day technical program was made up of presentations by recognized experts on a wide range of key topics including: • International Arctic oil and gas developments • Ice environments • Ice-going vessel technology • Remote sensing • Enhancements to mechanical recovery systems • Chemical herders in ice • Cold-water dispersants • Experimental spills • Case studies • Ongoing and future research programs
2007	Mechanical Recovery Systems for Ice-Infested Waters – Examination of Technologies for the Alaskan Beaufort Sea	A report identifying the existing state-of-technology for mechanical recovery in sea ice, and investigating any new mechanical recovery systems that may be transferable for use in ice-infested waters. More information is available at: http://dec.alaska.gov/spar/ipp/nscharter.htm
2007	Beaufort Sea Current Study	A report describing analyses designed to estimate the distance and direction an oceanic oil spill could travel during each of the two primary circulation regimes found in the nearshore Beaufort Sea: (1) under landfast ice during winter and (2) in open water or partial ice cover during summer.
		More information available at: http://dec.alaska.gov/spar/ipp/nscharter.htm
2008	Alaska In Situ Burning Guidelines	The Alaska in situ burning guidelines are used by the Alaska Department of Environmental Conservation, US Coast Guard, and US Environmental Protection Agency on-scene coordinators to authorize an emergency in situ burn of oil. More information is available at: http://dec.alaska.gov/spar/perp/docs/ISB-Rev1(Final-August%202008).pdf
2008	Ground Penetrating Radar (GPR) Workshop and Demonstration	A two-day GPR workshop was held on the North Slope to familiarize ADEC, ADNR, BPXA, ConocoPhillips, and ACS spill response personnel in using the GPR. ACS now has an operational GPR system on the North Slope.
2009–2010	Airborne GPR for Detecting Oil In and Under Ice	Continued studies of airborne supported ground penetrating radar (GPR) for detecting oil under sea ice. This project focuses on hardware development that will produce two prototype, higher-powered GPR systems that can be tested in Arctic field environments using commonly

Table B. (Continued)

YEAR	PROJECT TITLE	PROJECT DESCRIPTION
		available light helicopters. The goal is to significantly expand the practical operating window for oil detection on and under sea ice with GPR to cover a wider range of sea ice and climate conditions.
2009–2010	ADEC Tundra Treatment Manual	The initial document was produced in 2001. Additional research and studies are warranted to validate the procedures in the document, and documenting results of actual case studies of past spills that have impacted the tundra. Work includes field investigations and revising the guidelines, followed by changes to the manual, then production and distribution. More information is available at: http://dec.alaska.gov/spar/perp/r_d/ttman/tt_man.htm
2009–2010	Oleophillic Skimmer Research	In controlled conditions of cold and broken ice, test the effects on the Crucial disc skimmer's operation and durability and the skimmer's oil recovery rate and efficiency including the skimmer's ice-processing capability. In a test tank, measure the effects and recommend skimmer modifications to reduce ice interference in order to maximize recovery rate. Follow modified ASTM F2709 procedure as applicable to the conditions. These tests follow-up on Ohmsett tests that were conducted in non-Arctic conditions.
Ongoing	ShoreZone Mapping	ShoreZone is a mapping and classification system that specializes in the collection and interpretation of low-altitude aerial imagery of the coastal environment. Its objective is to produce an integrated, searchable inventory of geomorphic and biological features of the intertidal and nearshore zones, which can be used as a tool for science, education, management, and environmental hazard planning. Subareas with some if not entire coverage include Southeast Alaska, Prince William Sound, and Cook Inlet. For complete project description and status, visit: http://alaskafisheries.noaa.gov/habitat/shorezone/szintro.htm
		For other R&D projects, see: http://dec.alaska.gov/spar/perp/r_d/research_list.htm

In: Oil Spills in Arctic Waters
Editor: Nora Linwood

ISBN: 978-1-63321-580-1

Chapter 4

U.S. ARCTIC RESEARCH COMMISSION RECOMMENDS STEPS TO EXPANDED U.S. FUNDING FOR ARCTIC/SUBARCTIC OIL SPILL RESEARCH*

United States Arctic Research Commission

This paper was initially drafted prior to the Gulf Oil Spill and has since been modified to more accurately reflect concerns related to non-natural (anthropogenic) oil leaks and spills.

SUMMARY

The U.S. Arctic Research Commission (USARC) recommends steps that the U.S. government should take to invigorate oil spill research in the United States. The Commission specifically emphasizes the growing need for more-effective spill prevention and response in the Arctic region, and a greater understanding of oil and gas metabolism and toxicity in the Arctic marine ecosystem. If greater offshore energy exploration and production are realized, the risks of oil spills will increase concomitantly. The challenges of such exploration are compounded by the real and projected increase in Arctic shipping, by exploration in deeper, offshore waters, and by a marine environment that is undergoing dramatic change, particularly with respect to climate and ice conditions. The promise of a

* This is an edited, reformatted and augmented version of a white paper issued July 6, 2010.

rigorous and coordinated national research program on oil spills, made in the Oil Pollution Act of 1990 (OPA 90), after the *Exxon Valdez* disaster, has fallen short. The current spill-related disaster in the Gulf of Mexico makes this white paper all the more timely and should result in greater consideration of our proposed actions. USARC has worked closely with the federal Interagency Coordinating Committee on Oil Pollution Research (ICCOPR), stakeholders, and the public to develop ideas on this subject, which are integrated into this white paper.

This paper recommends several priority actions:

1) Government should update national and regional research plans as mandated by OPA 90 and should fund those plans, as authorized, through the $2.7 billion Oil Spill Liability Trust Fund (OSLTF), which is replenished by an eight cents per barrel tax on crude oil produced in or imported to the United States.1 ICCOPR is the leader of this federal research effort, as prescribed by law.
2) Expand the "endowment" funding for the National Oceanic and Atmospheric Administration (NOAA)-chaired Arctic/subarctic-focused spill research program created in OPA 90, the Prince William Sound Oil Spill Recovery Institute (OSRI). OSRI should serve as a funding model for the national ICCOPR program.
3) Increase National Science Foundation (NSF) and NOAA funding to enable scientists better understand the basic ecological structure and populations of key indicator species in the Arctic, including the deepwater column that feeds surface life. These species are important for subsistence and the survival of the entire ecosystem.
4) Increase understanding of the convective distribution of oil spilled in the ocean through additional support of physical oceanography studies (e.g., circulation, tides, and currents) of the Arctic Ocean and adjacent seas. There needs to be greater exploration and characterization of subsurface Arctic Ocean ecosystems to estimate plume lengths and distribution. Further develop, scale, and refine modeling efforts.
5) Study the metabolism of oil and gas (e.g., methane) and their toxicity to Arctic marine bacteria, benthos, and other members of the ecosystem. NOAA, NSF, and the Environmental Protection Agency (EPA) should be tasked accordingly.
6) Federal agencies should use a rigorous and thorough stakeholder consultation process in planning research and development objectives, funding them, and reviewing the results.
7) NOAA should co-chair ICCOPR, along with the U.S. Coast Guard (USCG), as NOAA has significant experience in directly conducting scientific research and in overseeing research conducted by NOAA-sponsored research entities. NSF should be

invited to join ICCOPR and, thus, become more engaged in oil spill research.

BACKGROUND

The United States acutely needs to improve capacity for oil spill research and response efforts in Arctic and Subarctic regions, including cases of both anthropogenic spills and leaks of oil and gas at surface and subsurface levels. Unique circumstances in the North include a virtually unexplored ocean ecosystem, protracted darkness, cold, variable ice conditions, and powerful storms. These conditions complicate prevention and response efforts for spills on land and sea, especially in ice-covered ocean waters. Basic biological considerations remain poorly known for the much of the region, such as the region's ecology and spatial habitat. We also need to better understand oil metabolism, toxicity, and accumulation in high-latitude flora and fauna (from the microbiological level on up). These data are necessary to develop appropriate impact assessment, mitigation strategies, and response intervention.

Despite these limitations, the Arctic is an area of increasing opportunity for both energy exploration and marine shipping. A 2008 U.S. Geological Survey (USGS) assessment[2] estimated that 13% of the world's undiscovered oil and 30% of its undiscovered gas lies within the Arctic. All five Arctic Ocean coastal states and Iceland have offshore exploration or production programs underway. There are currently over 600 active leases in Alaskan outer continental shelf (OCS) waters. A 2009 *Arctic Marine Shipping Assessment*[3], endorsed by the eight-nation Arctic Council, projects greater use of the Arctic Ocean by mariners. Over time, such shipping will initially serve communities and resource development, and may ultimately provide trans-Arctic "shortcut" routes for global shipping.

As the United States, Russia, Canada, Greenland, Iceland, and Norway all proceed with plans for high-Arctic oil and gas exploration, political and legal requests for improved spill response capability are on the rise. The current oil spill disaster in the Gulf of Mexico only amplifies the importance and the need for action. In 2007, North Slope Borough Mayor, Edward Itta, said, "We oppose offshore [drilling] until somebody proves to us they can clean up an oil spill in the Arctic"[4]. Though his views have tempered recently, the fear of oil spills impacting subsistence resources is very much alive in the mind of the public inhabiting a region in which any response capacity is minimal and fraught with difficulty. It is not surprising that representatives of the oil and gas industry argue that, with the appropriate precautions, the risks of oil spills are sufficiently low to enable safe, secure, and reliable exploration and production of offshore energy. Nevertheless, in communicating to USARC, these same industry representatives, in the United States, Canada, and Norway, express widespread agreement and support for increased research and development in spill prevention and response.

Federal research on Arctic ecosystems is primarily undertaken by NSF (Division of Polar Programs), NOAA, the Department of the Interior's Minerals Management Service (MMS), now renamed the Bureau of Ocean Energy (BOE), and the North Pacific Research Board. Federal research on Arctic oil spill prevention and response is currently carried out primarily by the USCG, BOE, NOAA's Coastal Response Research Center (CRRC), EPA, and OSRI. At least two federal interagency committees have been established by law to coordinate these activities: the Interagency Arctic Research Policy Committee (IARPC), established by the Arctic Research and Policy Act of 1984, which coordinates planning for all U.S. Arctic research programs throughout the government, and ICCOPR, established in OPA 90[5]. IARPC is chaired by NSF; ICCOPR is chaired by USCG.

Recent dialogue between USARC and ICCOPR has highlighted several "themes of importance." These themes relate to funding, Arctic-specific response (including defining the area in question, necessary baseline environmental research, unique conditions, such as ice-covered waters, and unique logistics), infrastructure capabilities, national priorities, opportunities for cooperation and collaboration between agencies and other stakeholder groups, use of "spills of opportunity" and controlled burns as research tools, building the next generation of researchers, and keeping abreast of industry research and development (R&D) reports with specific emphasis on the efforts of the International Petroleum Industry Environmental Association (IPIECA), which currently has a task force developing a prioritized list of additional research and technology projects to further advance oil spill preparedness in Arctic locations. We hope that, with the adoption of a regular ICCOPR meeting schedule, these themes will be addressed in crafting a new national research program.

USARC, under its authority to establish national policy, priorities, and goals for Arctic research, has long supported an appropriate basic and applied research program to find better methods to prevent and respond to oil spills in the Arctic region, as well as to understand the Arctic ecosystem and the metabolism and toxicity of oil released into the environment. We were asked by the State of Alaska to recommend that priorities meet natural resource management needs and the needs established by regulatory processes for contingency planning[6]. The Commission published *Oil Spill Response in Ice-Covered Waters* in 2004,[7] in which we found that "consistent long-term funding is needed for developing and improving response options for dealing with accidental oil spills in ice-covered waters." Following the 1989 *Exxon Valdez* oil spill and during the legislative consideration of the Oil Pollution Act of 1990, USARC supported the creation of the Oil Spill Recovery Institute[8]. USARC recently helped the United States complete the *Arctic Marine Shipping Assessment*,[9] in which the United States, along with the seven other Arctic nations, agreed on the need for more research[10]. The Commission also worked with Congress, the State of Alaska, and USCG to encourage an oil spill risk assessment in the Aleutian Islands. We co-sponsored a CRRC workshop[11] and a U.S.-Canada workshop[12], and visited the Joint Industry Program at SINTEF on oil in ice, which explored new spill mitigation strategies[13]. Finally, from a local perspective, the Commission recognizes and supports the call for in situ spill scenario testing promulgated by Mayor Itta and others. Integration of traditional knowledge into spill prevention and response efforts and oil spill-related social science research, essential to the determination of impacts, are also of great importance.

These priorities, and the actions that they support, are timely. If the debate on offshore exploration in America's Arctic, specifically the Beaufort and Chukchi seas, was not enough to make these proposals timely, the 2010 blowout in the Gulf of Mexico is. This disaster has the potential to cause severe environmental and economic effects in the southeastern United States and elsewhere, despite the fact that it happened in a temperate region with substantial and proximal spill response infrastructure available.

We hope now the United States will commit to funding a long-term, appropriate, and robust spill research program that contains a component that focuses on Arctic waters, including basic research on the biota, metabolism, and toxicity of oil in the Arctic. We urge the interagency committee to include a circumpolar approach in its research program to identify best management practices (BMPs)[14] from other nations' work in the Arctic, such as Norway's Barents Sea regulations. This approach will help ensure that research supports existing legal goals to constantly improve spill response. Alaska state and federal law requires spill prevention and response capabilities to improve dynamically over time[15], ensuring that best available technology and techniques are always deployed.

Despite the growing needs for such research, and the glaring absence of it as revealed by the ongoing Gulf of Mexico spill[16], much of the funding authorized in OPA 90 has expired. Oil pollution effects research, regional research programs, demonstration projects in New York and New Jersey, Los Angeles and Long Beach, and New Orleans, and a joint program from the Department of Commerce and EPA to monitor the environmental effects of oil discharges have all lost funding authorizations in the recent past. Although improvements are needed in both the ability to clean up oil spilled under ice and the detection of thin oil slicks trapped under ice in the Arctic and Subarctic regions, little progress has been made over the last two decades[17]. Recovery statistics for mechanical response techniques are similarly disappointing. Improvements are needed in areas such as health and human safety concerns (getting response personnel safely to spill sites), operability of equipment in icy arctic conditions, and rapid transport of equipment (including leak control devices) from populated areas with manufacturing capability (e.g., Houston) to remote spill sites. Concerns remain surrounding the environmental effects of in situ burning, chemical dispersants, and herding agents. Though some data exist, much additional research is needed in all of these areas, especially reegarding metabolism and toxicity on Arctic biota.

From the existing arctic/cold water response research, the "state of the art" is defined by technical reports and conference proceedings ("gray literature") that do not always meet the higher standards of peer review, and these documents are often not readily available to researchers and the public who might benefit from it. The Commission encourages ICCOPR to endorse the practice of publishing high-quality, peer-reviewed research related to oil spill response and prevention, as feedback has shown these types of articles are a necessary part of gaining the trust of stakeholder groups in research of this nature. We also suggest NSF/EPA foster the publication of results of oil metabolism and toxicity studies in Arctic biota, as well as studies involving modeling of convective plumes in the Arctic Ocean.

In a more favorable light, some research efforts are being conducted on priority issues, most recently by the SINTEF organization, in cooperation with various stakeholder groups. We encourage transparency in the conduct of research and peer-reviewed publication as a critical part of the broad distribution of the results. We look forward to learning more about the group's progress when the final report is released in the near future.

Despite these recent efforts, the Commission concludes that federal oil spill research efforts for Arctic conditions are fragmented, uncoordinated, under-funded, and in dire, immediate need of improvement.

Commission Recommendations

To this end, the Commission proposes the following coordinated actions among the Executive Branch, Congress, the State of Alaska and its municipalities, industry, academia, and other stakeholder groups:

1. *Strengthen interagency planning and coordination.* ICCOPR, created by OPA 90, should begin to meet regularly in a transparent fashion and with a regular agenda to develop justification for an appropriate level of national funding for oil spill research. Meetings should involve state environmental agencies, industry, and academic institutions, as they did in the early 1990s. ICCOPR needs to produce a regularly updated research and development program plan. It should be prioritized to reduce the greatest risks in the chain of oil exploration, production, transport, and use. Notices of meetings, minutes, and agendas should be posted online for the public to see. Congress should exercise its oversight, and the Office of Science and Technology Policy (OSTP) should exercise its coordination powers to ensure the research provisions of OPA 90 are followed. We recommend that ICCOPR establish a Federal Advisory Committee Act (FACA)-governed advisory committee comprised of a general advisory arm (involving academic and other nonfederal members) and a scientific advisory arm (involving academic and industry members, and others based on their scientific or technical expertise). This committee would help leverage public-private partnerships and promote stronger cooperation with nongovernmental researchers and may help build trust among stakeholders. With the expertise from advisory committee members, federal officials and the nation would have access to information and advice on a broad range of issues affecting ICCOPR policies and programs. The public, in return, would be afforded an opportunity to participate actively in the ICCOPR decision-making process.
2. *Create and fund a regional plan specific to the Arctic.* The Commission appreciates the fact that as we crafted this white paper, the members of ICCOPR met in Anchorage, Alaska, on March 4, 2010 with the Commission and Commission experts to review oil spill research needs in the Arctic. A day later, ICCOPR chair Capt. Anthony Lloyd joined the Commission in a public discussion of this

paper's recommendations, and on April 13, 2010, the Commission received additional written comments from ICCOPR.

The law sets out the means for research planning in OPA 90 and in the Arctic Research and Policy Act. Given the significant potential of offshore oil and gas in America's Arctic, an Alaska-specific research plan should be developed for presentation to both ICCOPR and IARPC. IARPC has already assigned the drafting of a "civil infrastructure research" plan to the U.S. Army Corps of Engineers, and an Arctic spill research plan could become part of that work product.

Regarding the development of an Alaska-region research plan for oil spill research, ICCOPR should work closely with other interagency research entities created by Congress and based in Alaska. One such entity is OSRI, which has already done much of the groundwork. OSRI provides public funding for Arctic/Subarctic spill research. The board of directors to OSRI, and their scientific advisory committee, includes representatives from a broad cross section of federal agencies, the State of Alaska, and the general public. The Commission met with the OSRI board in February 2010 to discuss this proposal. The matter is to be given further consideration at their upcoming meeting.

Other congressionally chartered research or spill response entities that should be involved include the members of the Regional Response Team for Alaska, the North Pacific Research Board, the North Slope Science Initiative, Regional Citizens Advisory Councils, the Arctic Institute of North America, and the Alaska Ocean Observing System. Additionally, Alaska coastal municipalities, the University of Alaska, and the University of New Hampshire's NOAA-sponsored Coastal Response Research Center have much to add. The plan should be cognizant of, and support cooperation with, oil spill research programs of Arctic industry and response cooperatives, including IPIECA, efforts of the Arctic Council, and well-established efforts at SINTEF in Norway and in the Arctic and Marine Oil-spill Program (AMOP) sponsored by Environment Canada. This list is not meant to be exclusive, but highlights existing infrastructure and cooperating groups that have made contributions so far. Senator Mark Begich has proposed legislation to fund research to further define Arctic Ocean research needs. Such a study should have an Arctic regional research plan to review. When a U.S. Arctic oil spill research plan is complete, it can be forwarded to ICCOPR and IARPC for inclusion in both national plans. We recommend plan renewal at least every five years.

3. *An "endowment approach" will ensure long-term funding.* Given recent lease sales earning close to $3 billion in revenues for the United States, other offshore development in Arctic/Subarctic ice-covered areas that will serve U.S. markets, and the increasing amount of shipping of all types occurring in the Arctic Ocean, USARC recommends a research funding level of $30 to $50 million per year for a national program, with $8–10 million per year dedicated to Alaska. This work would cover both the baseline biological research required in the Arctic as well as aggressively improve research, development, and on-water (in situ) experimentation of spill response in ice-covered conditions. Support is also needed for oiled wildlife response research and practices, as is additional funding specific to laboratory-based R&D for spill-response tactics. Major support should be given to basic research on Arctic ecosystems, including developing a thorough understanding of oil and gas metabolism and toxicity to Arctic microbes, benthic life, and other denizens of the water column, including fish and birds and mammals that depend on its productivity. NSF and EPA are likely to be the primary funding sources for these efforts. Oil spill research and development needs long-term funding continuity and commitment to facilities and people, particularly in the Arctic. An attached amendment to OPA 90 (Annex 1) would create a funding stream from the OSLTF18 and be spent through a competitive program managed by ICCOPR agencies. Besides the "endowment approach," Congress could authorize an appropriate amount of funding from the OSTLF to go to oil spill research by annual appropriation. The built-in "endowment approach" is now used to fund research and oil tank upgrade/replacement work by OSRI and the Denali Commission.
4. The Commission has considered the need for new legislation in the following areas:
   a) Support the aforementioned "endowment approach," and appropriate authorizations needed for the OSLTF to maintain a competitive research program involving industry and academic applicants with local stakeholders.
   b) Support an increase to the "endowment" fund for the OSRI by approximately $12 million for inflation proofing, as is now contained in Senate Bill S. 1194.

c) Support for Senator Mark Begich's initiative, Senate bill S. 1564, to fund a study to review research needs in the areas of spill response and prevention and to investigate the utility of Response Gap Analysis research. One example of this type of study would be a rapid report by the National Academy of Sciences. This report should include a focus on the metabolism and toxicity of oil in Arctic biota and include an analysis of novel biotechnology approaches to rapidly detoxify oil spills. We recommend the generation of a "State of the Science" report from these efforts to be compiled via a joint IARPC and ICCOPR effort followed by external peer review. This report would analyze existing information, define BMPs and use these data as a basis for defining new R&D priorities.
d) Expand the membership of the Interagency Oil Pollution Research Coordinating Committee to include NSF, OSRI, CRRC, industry, state, local and academic members, and tie ICCOPR's work to that of the White House-chaired National Science and Technology Council. Additionally, NOAA should co-chair ICCOPR, along with USCG, as NOAA has significant experience in directly conducting scientific research and in overseeing research conducted by NOAA-sponsored research entities.
e) Allow EPA and other appropriate regulators to waive restrictions that have so far prevented on-water testing of oil spills in the waters of the United States. Legislation could also encourage "spills of opportunity" to be used to test new response techniques.
f) Direct the Department of Justice to see that fines and penalties for oil spills are allocated to further support research.

Annex 1. Proposed Amendment to Increase Oil Spill Research Funding

The proposed amendment would fund the Oil Pollution Research and Development Program coordinated by the Interagency Coordinating Committee on Oil Pollution Research (33 U.S.C. 2761(c)) with annual interest earned on a $1 billion investment from the Oil Spill Liability Trust Fund established by 26 U.S.C. 2509. The purpose of this amendment is to create a dedicated funding stream for coordinated oil spill research without requiring new appropriations. It is important for the Committee to expand its research funding and coordination efforts as oil exploration, shipping, transportation, and other forms of commerce in Arctic waters increase in the coming years. The proposed amendment also modifies the existing statute to require agencies that are members of the Committee to spend Program funds on grants and cooperative agreements with independent entities including universities, research institutions, industry, and state and foreign governments to the maximum extent practicable.

1) Purpose and Awarding of Grants

Section 2761(c)(10) of title 33, United States Code, is amended by striking “may” from the first sentence and replacing it with “shall to the extent practicable” and by inserting “joint industry programs, pilot projects financed jointly with state or foreign governments,” before “and other persons.” Section 2761(c)(10) is further amended by inserting after the first paragraph the following:

“(a) Competitive awards. Contracts, cooperative agreements, and grants entered into under this section shall to the extent practicable be awarded on a competitive basis. Applications for awards will be subject to scientific merit review (peer review) and will be evaluated based on criteria developed by the Interagency Committee.”

2) Funding

Section 2761(f) of title 33, United States Code, is amended by striking paragraph (f) and inserting in lieu thereof:

“(f)(1) Amounts in the Fund shall be available without further appropriation and without fiscal year limitation, to carry out this section except for subsection (c)(8) of this section.

(2)USE OF INTEREST ONLY. The amount of funding to be made available annually to establish and implement the program under (c) of this section shall be the interest produced by the Fund’s investment of $1,000,000,000 and currently deposited in the Fund and invested by the Secretary of the Treasury in income producing securities along with other funds compromising the Fund. The National Pollution Funds Center shall transfer all such accrued interest annually to the Coast Guard beginning no more than six months after enactment of this Act, for the program.

(3) Congress may appropriate such additional funds as may be necessary to carry out this section."

End Notes

[1] 26 U.S.C. 4611 (c) (2) (B)

[2] Circum-Arctic Resource Appraisal: Estimates of Undiscovered Oil and Gas North of the Arctic Circle (2008). Available at: http://energy.usgs.gov/arctic/ (accessed July 12, 2010). See specifically "Slide Presentation."

[3] AMSA. 2009. Arctic Marine Shipping Assessment. Arctic Council, April 2009, second printing, 187 pp. Available online at: http://www.arctic.gov/publications/AMSA_2009_Report_2nd_print.pdf (accessed July 12, 2010).

[4] Ipsen, B. 2007. Residents voice opposition to Shell's offshore drilling. Pacific Environment, 19 April 2007. Available at: http://www.pacificenvironment.org/article.php?id=2340 (accessed July 12, 2010)/

[5] Oil Pollution Act of 1990, 101 H.R. 1465, P.L. 101-380. http://thomas.loc.gov/cgibin/bdquery/z?d101:h.r.01465:/.

[6] Alaska Department of Environmental Conservation, Division of Spill Prevention and Response, USARC Oil Spill White Paper comments, March 5, 2010.

[7] Oil Spill Response in Ice-Covered Waters (2004). Available online at: http://www.arctic.gov/publications/oil_in_ice.pdf (accessed July 12, 2010).

[8] Oil Spill Recovery Institute, http://www.pws-osri.org.

[9] AMSA. 2009. Arctic Marine Shipping Assessment. Arctic Council, April 2009, second printing, 187 pp. Available online at: http://www.arctic.gov/publications/ AMSA_2009_Report_2nd print.pdf (accessed July 12, 2010).

[10] The AMSA report is the result of a four-year, multinational-led project that was subsequently adopted in the 2009 Tromsø Declaration, a set of guidelines for the Arctic Council during the next two years that was ratified on April 29 by the eight Arctic states, including Deputy Secretary of State James Steinberg, who led the U.S delegation. Among many other findings, the AMSA report states that the "current lack of infrastructure" in the Arctic makes it more difficult to respond to spills because of the Arctic's "vast geographic distances in various seasonal and climactic circumstances."

[11] Opening the Arctic Seas: Envisioning Disasters and Framing Solutions. Report of a workshop held March 18–20, 2008, and sponsored by the Coastal Response Research Center at the University of New Hampshire. Available online at http://www.crrc.unh.edu/workshops/arctic_spill_summit/arctic_summit_report_final.pdf (accessed July 12, 2010).

[12] Northern Oil and Gas Research Forum: Current Status and Future Directions in the Beaufort Sea, North Slope and Mackenzie Delta. A workshop held in Anchorage, Alaska, October 28–30, 2009.

[13] The Joint Industry Program for Oil in Ice, Selskapet for Industriell og Teknisk Forskning ved norges tekniske hoegskole (The Foundation for Scientific and Industrial Research at the Norwegian Institute of Technology), http://www.sintef.no/Projectweb/JIP-Oil-In-Ice.

[14] Best management practices (BMPs) are defined under the Clean Water Act as a permit condition used in place of or in conjunction with effluent limitations to prevent or control the discharge of pollutants. BMPs may include a schedule of activities, prohibition of practices, maintenance procedure, or other management practice.

[15] Section 7001(c) of OPA 90 (33 U.S.C. sec. 2761) and Alaska Statutes 46.04.030(e).

[16] Schrope, M. 2010. Flood of oil, drought of research. Nature 27 465:404–405. Available online at: http://www.nature.com/news/2010/100525/full/465404a.html (accessed July 12, 2010).

[17] 2006 MMS Svalbard Experimental Spill to Study Spill Detection and Oil Behavior in Ice.

[18] The Oil Spill Liability Trust Fund was established by section 9509 of the Internal Revenue Code of 1986 (26 U.S.C. 9509).

In: Oil Spills in Arctic Waters
Editor: Nora Linwood
ISBN: 978-1-63321-580-1

Chapter 5

FEDERAL OIL AND GAS: INTERAGENCY COMMITTEE NEEDS TO BETTER COORDINATE RESEARCH ON OIL POLLUTION PREVENTION AND RESPONSE*

United States Government Accountability Office

WHY GAO DID THIS STUDY

Congress passed the Oil Pollution Act in 1990 (OPA). Among other things, OPA established the Interagency Coordinating Committee on Oil Pollution Research (interagency committee) to coordinate an oil pollution research program among federal agencies, including developing a plan, having the National Academy of Sciences review that plan, and reporting to Congress on the interagency committee's efforts biennially. The 2010 *Deepwater Horizon* explosion and fire led to the largest oil spill in U.S. history, raising new concerns about the effects of oil spills.

GAO was asked to assess the extent to which the interagency committee has facilitated the coordination of federal agencies' oil pollution research. (The Chairman, Subcommittee on Energy and Environment, House Committee on Science and Technology, now retired; and Representative Woolsey initiated

* This is an edited, reformatted and augmented version of a United States Government Accountability Office publication, No. GAO-11-319, dated March 2011.

this request.) In part, GAO analyzed committee documents and biennial reports and interviewed agency officials and nonfederal research entities.

WHAT GAO RECOMMENDS

GAO recommends, among other things, that the interagency committee coordinate efforts to evaluate the contributions of completed research and provide, in its 2012 biennial report to Congress, an update of its efforts to revise its research plan. The Department of Homeland Security concurred with our recommendations.

WHAT GAO FOUND

Federal agencies have conducted at least 144 research projects on oil pollution since 2003, but the interagency committee has played a limited role in coordinating this research, according to GAO's analysis of interagency committee reports and documents. For example, agencies conducted research on identifying the toxicity of nonpetroleum oils recovering oil from the sea floor. The interagency committee issued a research plan mandated by OPA in 1997 that set research priorities. This plan, however, did not fully address the recommendations on a draft plan made by the National Research Council, the organization through which the National Academy of Sciences provides most of its advice. For example, the National Research Council noted that the interagency committee should review and evaluate past and present oil pollution research to help guide federal efforts and to avoid duplication. The interagency committee has captured some member agencies' oil pollution research in its biennial reports to Congress, but it has not evaluated whether past research has advanced the 1997 research priorities; instead, the reports summarized projects. Without such an assessment, Congress may be less able to oversee the contributions of federal research to preventing and responding to oil spills. In addition, although OPA did not require that the interagency committee revise its 1997 plan, the National Research Council noted the need to continually reassess a comprehensive research plan. However, the interagency committee has not done so; consequently, the plan does not reflect changes in the oil production and transportation sectors since 1997, such as a significant increase in deepwater drilling. In September 2010, the interagency committee chair began to inventory completed research and categorize research projects according to the 1997 plan's research priorities, and the chair told GAO that the interagency committee will begin to update the 1997 plan in 2011.

OPA also directed the interagency committee to coordinate a comprehensive research program of oil pollution research among the member agencies, in cooperation with external stakeholders, such as industry, research institutions, state governments, and universities. An interagency member official told GAO that the committee helped foster interagency cooperation between two agencies comparing two types of testing to determine the effectiveness of certain chemicals in dispersing oil in sea water; However, more generally, the interagency committee took limited action to foster communication among member agencies between 1997 and 2009, when the chair proposed updating the 1997 plan, according to some member agency officials. Although the interagency committee's meetings have occurred once or twice annually for the past 2 years, they occurred irregularly before then. Additionally, member agencies were not consistently represented in the interagency committee. In October 2010, to better communicate with interagency committee member agencies, among others, the interagency committee launched a Web site, which provides transcripts from its past public meetings and biennial reports to Congress.

ABBREVIATIONS

BOEMRE	Bureau of Ocean Energy Management, Regulation and Enforcement
EPA	Environmental Protection Agency
NASA	National Aeronautics and Space Administration
NOAA	National Oceanic and Atmospheric Administration
NRT	National Response Team
OPA	Oil Pollution Act of 1990
OSRI	Oil Spill Recovery Institute
PHMSA	Pipeline and Hazardous Materials Safety Administration

March 25, 2011

The Honorable Lynn Woolsey
House of Representatives

Dear Ms. Woolsey:

On April 20, 2010, an explosion and fire onboard the *Deepwater Horizon* drilling rig in the Gulf of Mexico led to the largest oil spill in U.S. waters. The total cost of cleaning up this massive spill, the extent of the damage to the environment, and the potential effect on the Gulf Coast states' economies will not be known for some time. However, current estimates suggest that spill cleanup and related damage claims will cost tens of billions of dollars—far more than the more than $2 billion that was spent to clean up the 1989 *Exxon Valdez* spill, which contaminated Alaska's south central coastline, including portions of national wildlife refuges, national and state parks, a national forest, and a state game sanctuary.

Following the *Exxon Valdez* spill, Congress passed the Oil Pollution Act of 1990 (OPA). Among other things, OPA addresses liability for the costs of cleaning up spills and damages to the environment. OPA also established the Interagency Coordinating Committee on Oil Pollution Research (the interagency committee) to coordinate a comprehensive program of oil pollution research, technology, development, and demonstration among federal agencies. OPA, as amended, requires the U.S. Coast Guard to chair the 13-member interagency committee. In addition, the chair is required to report every 2 years to Congress on the committee's past activities and future plans for oil pollution research.[1] OPA also directed the interagency committee to develop a comprehensive research and technology plan to lead federal oil pollution research. In response to this directive, the interagency committee issued a plan in 1997 to guide research on oil pollution prevention and response. Additionally, OPA authorized funds from the Oil Spill Liability Trust Fund to, among other things, pay for certain oil pollution research.

In this context, you asked us to review the interagency committee's work concerning federal oil pollution research. (This request was originally made by the Chairman, Subcommittee on Energy and Environment, House Committee on Science and Technology, now retired; and Representative Woolsey.) Our objective was to assess the extent to which the interagency committee has facilitated the coordination of federal agencies' oil pollution research. To address this objective, we reviewed OPA to understand the interagency committee's purpose and charge. We reviewed the interagency committee's biennial reports to Congress for the fiscal years 2000 through 2009 to assess efforts to identify and set priorities for research needs. We focused on the period since 2000 because of Congress's interest in the interagency committee's current and recent activities. We interviewed cognizant agency officials on the extent of coordination among interagency committee member agencies and, in September 2010, attended a public meeting of the interagency committee to observe efforts to coordinate oil pollution research. We also obtained the views of stakeholders, such as state agencies and a nonprofit research organization. We reviewed interagency committee documentation and the National Research Council's report on the interagency committee's research plan to assess the committee's efforts to evaluate research projects and determine progress made toward completing research goals. We reviewed committee documentation and interviewed cognizant agency officials about any current and emerging oil pollution risks and discussed how they were identified. To determine the number of research projects conducted by member agencies, we reviewed the interagency committee's biennial reports to Congress. We intended to identify the number of projects conducted since the completion of the 1997 research plan but could not do so for fiscal years 1997 through 2002 because research projects were not reported separately during those years. Because of concerns about the availability and reliability of data, we were not able to identify all research projects completed during those years; however, we believe we captured the majority of the projects with our methodology.

We conducted this performance audit from June 2010 to March 2011 in accordance with generally accepted government auditing standards. These standards require that we plan and perform the audit to obtain sufficient, appropriate evidence to provide a reasonable basis for our findings and conclusions based on our audit objectives. We believe that the evidence obtained provides a reasonable basis for our findings and conclusions based on our audit objectives. A further discussion of our scope and methodology is presented in appendix I.

BACKGROUND

This section provides information on OPA requirements, expenditures for oil pollution research conducted by interagency committee member agencies, and certain other organizations that conduct or coordinate research.

The Interagency Committee on Oil Pollution Research

Through OPA Congress established the interagency committee to coordinate a comprehensive oil pollution research program among federal agencies and in cooperation with industry, universities, research institutions, state governments, and other nations, as appropriate. It also designated member agencies, authorized the President to designate other federal agencies, and directed that a representative of the Coast Guard chair the interagency committee.

The chairman's duties include reporting biennially to Congress on the interagency committee's member agencies' activities related to oil pollution research.

As also directed by OPA, the interagency committee was to develop a research plan that:

- identified member agencies' roles and responsibilities;
- assessed the current status of knowledge on oil pollution prevention, response and mitigation technologies, and effects of oil pollution on the environment;
- identified significant oil pollution research gaps, including an assessment of major technological deficiencies in responses to past oil discharges;
- established research priorities and goals for oil pollution technology development related to prevention, response, mitigation, and environmental effects;
- estimated the resources needed for federal agencies to conduct the oil pollution research and development program and timetables for completing research tasks; and
- identified, in consultation with the states, regional oil pollution research needs and priorities for a coordinated, multidisciplinary program of research at the regional level.

OPA also directed the chair of the interagency committee to contract with the National Academy of Sciences to (1) provide advice and guidance in the preparation and development of the research plan and (2) assess the adequacy of the plan as submitted and submit a report to Congress on the conclusions of that assessment.[2]

The interagency committee prepared the original research plan and, in 1992, submitted it to Congress and the National Research Council—created under the auspices of the National Academy of Sciences and through which the academy provides most of its advice—for their review and comment. The second edition of the research plan was submitted to Congress on April 1, 1997.

Interagency Committee Member Agencies' Expenditures for Oil Pollution Research

According to agency officials, since fiscal year 2000, member agencies have spent about $163 million on oil pollution research. Of this total, approximately $145 million came from the Oil Spill Liability Trust Fund authorized by OPA.[3]

The largest source of revenue for the trust fund has been a tax collected from the oil industry on petroleum produced in or imported into the United States. The tax, which was $0.05 per barrel when OPA was enacted, expired in 1994 but was reinstated in 2005 and increased to $0.08 per barrel in 2008.

Member agencies spent an additional $18 million on oil pollution. Table 1 shows the sources of funding for oil pollution research among seven interagency committee member agencies who reported that they conducted oil pollution research: the Bureau of Ocean Energy Management, Regulation and Enforcement (BOEMRE); the Coast Guard; the Environmental Protection Agency (EPA); the National Aeronautics and Space Administration (NASA); the U.S. Navy; the National Oceanic and Atmospheric Administration (NOAA); and the Pipeline and Hazardous Materials Safety Administration (PHMSA).[4]

Other Organizations that Conduct or Coordinate Oil Pollution Research

After the 1989 *Exxon Valdez* spill in Prince William Sound, Alaska, at least four states created or expanded their own oil pollution research programs and Congress created an oil pollution research institute.

Table 1. Interagency Committee Member Agencies' Funding for Oil Pollution Research, Both Trust and Agency Funds, Fiscal Years 2000 through 2010

Fiscal year 2010 dollars in millions												
	Fiscal year											Agency
Member agency	**2000**	**2001**	**2002**	**2003**	**2004**	**2005**	**2006**	**2007**	**2008**	**2009**	**2010**	**total**
BOEMRE[a]	$7.1	$6.9	$6.8	$6.7	$7.5	$7.9	$7.5	$7.5	$7.1	$6.6	$6.2	**$77.8**
Coast Guard	4.7	4.8	4.8	4.3	4.1	2.3	2.8	2.1	2.1	0.6	0.7	**$33.3**
EPA	1.1	1.1	1.1	1.1	1.0	1.0	0.9	0.8	0.9	0.7	0.6	**$10.3**
NASA	0	0	0	0	0	0	0	0	0.4	0	0	**$0.4**
Navy	4.3	3.5	1.2	0.4	0.3	0.3	0.2	0.3	0.6	0.4	0.6	**$12.1**
NOAA	0	0	0	0	2.3	2.2	3.3	3.2	0	0	0	**$11.0**
PHMSA	0	0	0	0	3.4	3.3	2.1	2.0	3.4	2.2	2.2	**$18.6**
Annual total	**$17.2**	**$16.3**	**$13.9**	**$12.5**	**$18.6**	**$17.0**	**$16.8**	**$15.9**	**$14.5**	**$10.5**	**$10.3**	**$163.5**

Source: GAO analysis of data provided by interagency committee member agencies.

[a]Expenditures for BOEMRE include about $3 million per year for the operation and maintenance of the National Oil Spill Response Research and Renewable Energy Test Facility located in Leonardo, New Jersey. This facility is used for both full scale equipment testing and responder training.

- *Alaska Division of Spill Prevention and Response*. This division was established in 1991, although an official from the Alaska Division of Spill Prevention and Response told us that the state has had an oil pollution control program, which included research, since the 1970s. According to the agency's Web site, Alaska appropriated a total of $2.5 million in the wake of the *Exxon Valdez* oil spill to enhance the ability of the state and industry to respond to oil spills. The funds were to be used for research programs directed toward the prevention, containment, cleanup, and amelioration of oil spills in Alaska. To date, more than 30 research and development projects have been completed.
- *California Office of Spill Prevention and Response*. This office was created in 1990 and has a variety of responsibilities related to spill prevention and response, including oil spill contingency planning. The office's research program operated from 2004 through 2010 and supported a total of 38 research projects with a budget of $430,000 annually during this 6-year period.
- *Louisiana Applied and Educational Oil Spill Research and Development Program*. Louisiana's program was established after the *Exxon Valdez* oil spill. The state created the Louisiana Oil Spill Coordinator's Office, which, with Louisiana State University, formed the Oil Spill Research and Development Program. The program's mission was to provide the state of Louisiana with tools related to oil spill prevention, detection, response, and cleanup. According to a program official, from 1993 through 2007, the program provided more than $500,000 per year to public colleges and universities to support a range of research.
- *Texas General Land Office Oil Spill Prevention and Response Program*. According to a state official, the Texas General Land Office's Oil Spill Prevention and Response Program has spent $1.25 million per year for oil spill research since 1991. Its research is funded by a fee on oil loaded or unloaded in Texas.
- *Oil Spill Recovery Institute (OSRI)*. OPA established OSRI for research, education, and demonstration projects to respond to and understand the effects of oil spills in the Arctic and sub-Arctic marine environments, amongst other purposes. OSRI is administered through and housed at the Prince William Sound Science Center, a nonprofit research and education organization in Cordova, Alaska. Funding for OSRI comes from interest on $22.5 million in the Oil Spill Liability

Trust Fund. OSRI received more than $1 million from the fund in 2009 and $225,000 in 2010 and expects to receive between $560,000 and $1.3 million in 2011, according to an agency official.

In addition, the National Response Team (NRT) coordinates some oil pollution research. NRT is an interagency organization responsible for, among other things, coordinating emergency preparedness and response to oil and hazardous substance pollution incidents.[5] EPA and the Coast Guard serve as its Chair and Vice Chair, respectively. One of NRT's responsibilities is to monitor "response related research and development, testing and evaluation activities of NRT agencies to enhance coordination, avoid duplication of effort and facilitate research in support of response activities." Every 2 years NRT's science and technology committee—which includes, among others, BOEMRE, the Coast Guard, EPA, and NOAA— provides the interagency committee with the information for its biennial reports to Congress. The science and technology committee also meets monthly and member agencies coordinate regularly on oil pollution research projects. These meetings allow agencies to leverage each other's resources to achieve mutually beneficial oil pollution research, according to agency officials.

Federal Agencies Have Conducted Oil Pollution Research, but with a Limited Coordination Role by the Interagency Committee

According to our analysis of interagency committee reports, federal agencies have conducted at least 144 research projects on oil pollution prevention and response since 2003, but the interagency committee had a limited role in facilitating the coordination of agency efforts.[6] The interagency committee established a joint research plan in 1997 that identified oil pollution risks and research priorities, but it has not updated that plan in light of changes in the oil production and transportation sector. The interagency committee also submitted biennial reports to Congress, as directed, but it has not evaluated member agencies' progress in addressing research gaps identified in the 1997 research plan; until recently, it also had not revisited the plan, as the National Research Council recommended. Furthermore, since completing the 1997 research plan, the interagency committee has taken limited action, until recently, to foster communication and coordinate research among member agencies and to reach out to stakeholders, such as industry and state organizations.

Federal Agencies Have Conducted at Least 144 Research Projects on Oil Pollution Prevention and Response since Completion of the Research Plan

According to the interagency committee's biennial reports, since 2003 member agencies have conducted at least 144 research projects related to preventing or responding to oil pollution. These projects have addressed a range of topics, such as responding to an oil spill by burning oil off the water's surface (*in situ* burning), detecting oil in icy waters, predicting oil behavior in deepwater blowouts, and using micro-organisms to remove spilled oil in saltwater marshes. As table 2 shows, BOEMRE, the Coast Guard, EPA, and NOAA—4 of the 13 member agencies—accounted for all of the projects reported to Congress. Of the remaining nine member agencies, three agencies conducted research, but their research was not reported in the interagency committee's biennial reports, and six agencies did not conduct any research.

Table 2. Number of Oil Pollution Research Projects Conducted by Member Agencies as Reported in the Interagency Committee's Biennial Reports to Congress, Fiscal Years 2003 through 2010

	Member agency					
Fiscal year	BOEMRE	Coast Guard	EPA	NOAA	Joint	Total
2003	12	1	12	0	0	25
2004	8	1	9	0	0	18
2005	8	0	8	2	0	18
2006	7	1	5	0	1[a]	14
2007	7	1	3	10	1[b]	22
2008	3	0	2	8	1[c]	14
2009	9	2	2	9	0	22
2010	2	0	5	4	0	11
Total	56	6	46	33	3	144

Source: GAO analysis of interagency committee biennial reports to Congress.

Notes: We could not identify the number of projects completed in 2000 because the interagency committee was not required to report on its progress that year per the Federal Reports Elimination and Sunset Act of 1995, and did not do so. Additionally, we could not identify the number of projects completed in fiscal years 2001 and 2002 because the interagency committee's biennial reports did not include projects; instead, the reports included publications authored by member agencies, and we could not confirm whether individual publications corresponded to a single project. The interagency committee plans to release the next biennial report in 2012, which will cover 2010 and 2011.

[a] In 2006, BOEMRE and EPA jointly conducted a research project to evaluate the reliability of a laboratory method used to test dispersant effectiveness.

[b] In 2007, the Coast Guard and NOAA jointly developed a tool to help responders and planners assess the risk from chemical spills in rivers.

[c] in 2008, BOEMRE and the Coast Guard completed a joint research project on verifying the reliability of a cooperatively designed monitoring program to use when *in situ* burning occurs and dispersants are used.

Projects conducted by these agencies and included in the interagency committee's biennial reports addressed a wide range of topics. For example:

- *BOEMRE: research to develop an aerial oil thickness and mapping system.* Based on this research, initiated in 2005, BOEMRE developed a portable aerial sensor to detect and accurately map the thickness and distribution of oil slicks in coastal and offshore waters. The aerial thickness mapping system was deployed for the *Deepwater Horizon* oil spill and flown over the spill, providing maps of oil thickness. The Coast Guard used these maps to guide mechanical response efforts and dispersant operations and to plan *in situ* burns, according to Coast Guard officials. In addition, NOAA used this information to validate its model predictions for how the oil would behave in water, to document the potential for the oil to arrive on beaches, and to assess oil infiltration to the shoreline and marshes, according to NOAA officials.
- *Coast Guard: recovery of oil on the sea floor.* This project, which is ongoing is intended to develop methods to recover oil located on the bottom of the sea, according to Coast Guard officials. Its first objective is to develop a number of potential methods for detecting the oil and then selecting the most cost effective methods for further development.
- *EPA: research into the biodegradability and toxicity of nonpetroleum oils.*[7] Through its ongoing research, EPA has found that the degree to which vegetable oils will biodegrade in the environment depends on a number of factors, including the oil's chemical structure, according to EPA officials. Also, EPA found that vegetable oils can readily

biodegrade anaerobically—or without oxygen—suggesting that a new treatment technology could be used for cleaning up a vegetable oil spill. This technology involves sinking the oil into the sediment by adding clay so that the oil rapidly biodegrades under anaerobic conditions with little adverse effects on the ecosystem. Currently, the National Contingency Plan provides that sinking agents may not be used as an oil recovery or mitigation measure, but as a result of this research, EPA is considering proposing an exception for treating vegetable oil spills.

- *NOAA: research into monitoring the effectiveness of chemicals used to disperse oil.* This research, completed in 2008, compared the behavior of oils with and without dispersants in different types of sediment from U.S. coastal waters, according to the interagency committee's 2008–2009 biennial report.

While these four agencies' research projects were discussed in the interagency committee's biennial reports, three other member agencies also conducted research that was not reported, according to our analysis of information that some agencies provided. In speaking with agency officials, however, we could not determine why the following agencies were omitted from the interagency committee's biennial reports.

- PHMSA has administered an oil pollution research program since fiscal year 2002, but none of its projects have been included in the biennial reports. For example, PHMSA has an ongoing project to develop a model for commercial companies to predict the rate at which operating pipelines become weakened and suddenly fracture because of stress and corrosion, and in 2009, PHMSA completed a project examining the risk of plastic pipe failures, according to PHMSA documentation.
- The Navy and NASA have conducted some oil pollution research, but none of their research efforts were included in the biennial reports. For example, the Navy has an ongoing, multiphase project to evaluate the efficacy of equipment used to separate oil from wastewater before the wastewater is discharged from Navy ships. The Navy decided to research this issue because the chemical and physical properties of synthetic lubricants, some of which are denser than water, have posed problems for its oil-water separators, which operate based on the differences in specific gravity between oil and water, according to Navy documentation. Similarly, NASA recently provided funding to

an oil pollution detection project through its Gulf of Mexico Initiative.[8] The goal of the project, which is being conducted in partnership with the Naval Research Laboratory and NOAA, is to demonstrate practical applications for oil spill detection from observations of two NASA sensors in low-earth orbit. From these observations, NASA officials said that new methods will be developed for NOAA to use to detect oil spills. NASA officials said they selected this project because it would employ an innovative use of remote sensing technology, not because of its focus on detecting oil spills.[9]

Without knowing about these projects, Congress may be less informed when making funding decisions about oil pollution research.

The Interagency Committee Coordinated Efforts to Develop the 1997 Research Plan, but until 2009, Took Limited Action to Foster Communication and Coordinate Research

The interagency committee completed the research plan mandated by OPA to help guide member agencies' research on oil pollution prevention and response in 1997. However, once the plan was completed, the interagency committee played a limited role in coordinating member agencies' efforts.

The Interagency Committee Developed the 1997 Research Plan through Joint Efforts but Has Not Addressed Some National Research Council Recommendations

The interagency committee prepared a research plan required by OPA and submitted it for review to the National Research Council and Congress in 1992. The National Research Council provided its review of the first plan in 1994, and the interagency committee submitted the second edition of the plan to Congress on April 1, 1997. According to the interagency committee's documentation, the committee conducted a 2-year voluntary interagency effort to address the National Research Council's recommendations. The interagency committee's 1997 research plan includes (1) an analysis of the oil production and transportation systems and associated oil pollution risks; (2) an identification of 21 research priorities intended to address oil pollution risks, categorized into three priority levels; (3) an identification of research areas of focus for some member agencies; and (4) an identification of some nonfederal stakeholders.

While the interagency committee revised its research plan in order to address the National Research Council's review, the committee did not fully address all of the council's recommendations. For example, after reviewing the interagency committee's first draft research plan, the National Research Council noted the interagency committee should, as part of its activities, comprehensively review and evaluate past and present oil pollution research to help guide federal research efforts and avoid duplication. The interagency committee followed this recommendation, in part, by capturing the results of some member agencies' oil pollution research in its biennial reports to Congress, but it did not assess whether completed research contributed to advancing the 1997 research priorities; rather, the reports provided only summaries of research projects. Without such an assessment, Congress may be less able to provide oversight on the contributions of federal research to prevent and respond to oil spills. Furthermore, while some member agencies maintain Web sites that are accessible to the public and that contain data and reports on oil pollution research that has been conducted, the interagency committee has not assembled or published a comprehensive inventory of all research projects conducted by member agencies, which limits the interagency committee's ability to evaluate past research.

The interagency committee has recently taken steps to inventory member agencies' research. Specifically, according to Coast Guard documents, in September 2010, the interagency committee chair began to inventory research projects and categorize them according to the 1997 plan's research priorities. The interagency committee chair told us that this inventory is likely to help the interagency committee determine where to focus future research efforts in response to current and emerging risks.

In addition, while OPA did not require the interagency committee to revise its research and technology plan, the National Research Council noted in its review that a comprehensive research plan should be continually reassessed. However, the interagency committee has not revised its 1997 research plan. As a result, the plan does not reflect significant changes in the oil production and transportation sectors or assess current and emerging risks or research priorities. Consequently, knowledge gaps in critical research areas may have been overlooked. For example:

- The 1997 plan contained 21 research priorities, such as oil spill surveillance and environmental restoration methods, and identified knowledge gaps in these areas, but it did not identify deepwater drilling as a specific research priority. However, by 2000, deepwater

oil production had surpassed shallow water oil production, and within 5 years of the plan's completion, oil production in deepwater had tripled, according to data from BOEMRE.[10]
- The plan did not identify oil spills in icy waters as a risk, although oil production and shipping are expected to increase substantially in the Arctic, according to member agency officials.

Coast Guard officials said that although the 1997 plan did not focus on oil spills in deepwater or the Arctic, many of the plan's research priorities are still relevant for guiding current research. However, most officials from the 13 member agencies we spoke with told us that they either did not know that the interagency committee's 1997 plan existed or did not use it to guide research; rather, each agency determined its own research priorities based on its mission. For example, EPA used a multiyear plan to guide all of its research, including oil pollution, but its plan did not reference the interagency committee's 1997 research plan.

Recognizing the need for a more active approach, the interagency committee chair told us that the committee began to consider updating the 1997 plan in late 2009 and planned to ask member agency officials to draft components of the revised plan during the summer of 2010. However, a number of member agencies were occupied with responding to the *Deepwater Horizon* incident, according to agency officials, and were thus unable to begin revising the plan. Coast Guard officials expect drafting of a revised research plan to begin during the summer 2011 and stated that it will take approximately 2 years to update the plan because the interagency committee intends to submit the plan to the National Research Council for its review. Coast Guard officials said that this effort to review and revise could take several years, as it did in the 1990s. Furthermore, according to Coast Guard officials, they have not yet decided whether the new research plan will include an evaluation of past research or address research priorities outlined in the 1997 plan.

Interagency Committee Has Taken Limited Actions to Foster Communication and Coordination among Member Agencies and Nonfederal Stakeholders

As directed by OPA, the interagency committee was to coordinate a comprehensive program of oil pollution research among the member agencies, in cooperation and coordination with industry, universities, research institutions, state governments, and other nations, as appropriate. The interagency committee has helped member agencies collaborate on some occasions. For example, according to an agency official who participates in the interagency committee, the committee played a role in facilitating interagency cooperation between BOEMRE and EPA. These agencies jointly conducted research, completed in 2006, in comparing how laboratory tests of the effectiveness of certain chemicals in dispersing oil in sea water compared with certain larger scale tests at a research facility.

According to some member agency officials, however, the interagency committee had taken limited action to foster communication among member agencies between 1997 and 2009, when the interagency committee chair proposed updating the 1997 plan. Although the interagency committee's meetings have occurred once or twice annually for the past 2 years, they occurred irregularly before then, according to some agency officials.

Additionally, member agencies were not consistently represented in the interagency committee. Specifically, five agencies did not have a representative designated to the interagency committee until 2010. An official at one of these agencies told us that he was assigned as the representative to the interagency committee only after the agency had received our request to discuss the interagency committee's work. Furthermore, officials at one agency said that they have never heard of the interagency committee and reported that the agency did not have a representative designated to the interagency committee.

In October 2010, to better communicate with interagency committee member agencies, among others, the Coast Guard launched the interagency committee's Web site, which includes transcripts from past public meetings and biennial reports to Congress. In addition, as directed by OPA, the interagency committee was to cooperate and coordinate with industry, universities, research institutions, state governments, and other nations, as appropriate. With specific regard to states, the interagency committee was to consult with them on regional oil pollution research needs and priorities. The National Research Council echoed these requirements in its recommendations, noting that such work was necessary in order to avoid duplication of research efforts and to enhance coordination and cooperation with those entities. In its 1997 research plan, the interagency committee identified the activities of some stakeholders, including the oil pollution research programs of four states and three industry groups, but interested stakeholders have reported limited contact with the interagency committee. For example:

- Officials from two of the four state oil pollution research programs we spoke with were unaware of the interagency committee's existence until we contacted them.
- Officials from the other two state oil pollution research programs reported having past, albeit inconsistent, interaction with the interagency committee.
- The committee hosted three public meetings in 2010 to solicit input from nonfederal stakeholders on the direction of a new research plan; however, it announced the meetings only 4 weeks in advance, which may have been insufficient time to obtain participation from a range of stakeholders.
- An official we spoke with from a nonprofit oil pollution research organization had never interacted with the interagency committee until two of the conferences in 2010.

By not communicating with key nonfederal stakeholders, the interagency committee may have missed opportunities to coordinate research efforts across sectors. For example, a state official we spoke with said that he is concerned that the interagency committee is not doing a sufficient job to minimize the duplication of research efforts across sectors; he noted that some of the federal and state research recently completed or currently underway is similar to federal and state research completed in the 1990s. Several state officials we spoke with also said that the interagency committee has generally not done a sufficient job of disseminating the results of completed federal research to nonfederal stakeholders, which could help nonfederal research organizations in planning their own research efforts. Furthermore, while the interagency committee's last biennial report listed workshops or conferences interagency members attended, it did not report on any efforts to consult with key nonfederal stakeholders.

In December 2010, Coast Guard officials told us that the interagency committee was considering establishing a subcommittee to coordinate with industry on planning and research, but they had not yet firmed up any plans to do so.

CONCLUSION

Like the *Exxon Valdez* spill in 1989, the *Deepwater Horizon* incident once again highlighted the need for new knowledge about oil spill prevention and response. The interagency committee completed a research plan required by OPA in 1997 to help guide member agencies' research on oil pollution prevention and response. Federal agencies have conducted at least 144 research projects related to this issue, but the interagency committee, established to develop a comprehensive research and development program on oil spill prevention and response, has been incomplete in its accounting for research projects and has done little until recently to coordinate the federal research effort.

The chair of the interagency committee has recognized the need for a proactive approach to coordination, and the committee's recent effort to inventory member agencies' research projects is a necessary step to understanding past research. However, this effort will be incomplete without an evaluation of whether this research addressed knowledge gaps identified in the 1997 plan. Without such an evaluation, Congress may be unable to provide effective oversight on the progress made in federal efforts to conduct research on oil pollution prevention and response. Furthermore, Coast Guard officials expect the drafting of a revised research plan to begin during summer 2011, but the revision of the plan has already been delayed because of the *Deepwater Horizon* incident, and the interagency committee could take several years to complete the planned revision, as it did in the 1990s with the 1997 research plan. Moreover, in the past, the interagency committee has not reached out effectively to identify and consult with key nonfederal stakeholders who could provide insight into the research that may need to be conducted, as it was directed to do by OPA. Without such outreach, the committee may be missing opportunities to advance knowledge across sectors and to avoid duplication of research efforts.

RECOMMENDATIONS FOR EXECUTIVE ACTION

In order to better identify oil pollution risks, determine research priorities, and coordinate research efforts, we recommend that the Commandant of the U.S. Coast Guard direct the chair of the interagency committee to take the following three actions, in coordination with member agencies:

- Evaluate the contributions of past research to current knowledge on oil pollution prevention and response and report the results of these evaluations, including remaining gaps in knowledge, in its biennial reports to Congress.

- Provide a status update regarding the revision of the research plan, as well as a schedule for completing the revision, in the next biennial report due in 2012, which will cover 2010 and 2011.
- Establish a more systematic process to identify and consult with key nonfederal stakeholders on oil pollution risks and research needs on an ongoing basis.

AGENCY COMMENTS AND OUR EVALUATION

We provided the departments of Commerce, Defense, Energy, Homeland Security, the Interior, and Transportation; EPA; and NASA with a draft of this report for review and comment. In commenting on this report, the departments of the Interior and Transportation, and EPA provided technical comments, which we incorporated as appropriate. In addition, the Department of Homeland Security concurred with our recommendations.

Sincerely yours,
Frank Rusco
Director, Natural Resources and Environment

APPENDIX I: SCOPE AND METHODOLOGY

To review the extent to which the Interagency Coordinating Committee on Oil Pollution Research (interagency committee) has facilitated the coordination of federal agencies' oil pollution research efforts, we analyzed biennial reports produced by the interagency committee to assess efforts to identify and set priorities for research needs and reviewed our guidance on interagency collaboration.[1] We interviewed cognizant agency officials on the extent of coordination among committee member agencies and, in September 2010, we attended a public meeting of the interagency committee to observe efforts to coordinate oil pollution research. We also interviewed external stakeholders, including officials from California, Louisiana, and Texas, and the Oil Spill Recovery Institute, a nonprofit research organization. We selected these organizations because all were listed in the interagency committee's research plan as stakeholders. The findings from the officials we interviewed, however, cannot be generalized to other states or organizations. We also reviewed and analyzed interagency committee documentation to assess efforts to evaluate research projects and determine progress made toward completing research goals. We reviewed committee documentation and interviewed cognizant agency officials about any current and emerging oil pollution risks, as well as how they were identified.

To determine the number of research projects conducted by member agencies, we reviewed the interagency committee's biennial reports to Congress. While we intended to count the number of projects conducted since completion of the 1997 research plan, we could not count projects for fiscal years (1) 1997 and 1998 because the biennial report that includes those years did not include any research projects initiated after completion of the research plan; (2) 1999 and 2000 because the interagency committee was not required to report on its progress for those two years in accordance with the Federal Reports Elimination and Sunset Act of 1995, and did not do so; and (3) 2000, 2001, and 2002 because the interagency committee's biennial reports included publications and not projects. Also, we could not confirm whether individual publications corresponded to a single project. Because of concerns about the availability and reliability of data, we were not able to identify all research projects completed during those years; however, we believe we captured the majority of the projects with our methodology because we were able to interview program officials from each member agency that conducted oil pollution research and confirm our approach and our list of projects with them.

We conducted this performance audit from June 2010 to March 2011 in accordance with generally accepted government auditing standards. These standards require that we plan and perform the audit to obtain sufficient, appropriate evidence to provide a reasonable basis for our findings and conclusions based on our audit objectives. We believe that the evidence obtained provides a reasonable basis for our findings and conclusions based on our audit objectives.

End Notes

[1] The agencies assigned to the interagency committee under OPA include the Department of Commerce's National Oceanic and Atmospheric Administration and National Institute of Standards and Technology; Department of Defense's U.S. Army Corps of Engineers and U.S. Navy; Department of Energy; Department of Homeland Security's U.S. Coast Guard and Federal Emergency Management Agency; Department of the Interior's Minerals Management Service (now the Bureau of Ocean Energy Management, Regulation and Enforcement) and U.S. Fish and Wildlife Service; Department of Transportation's Maritime

Administration and Pipeline and Hazardous Materials Safety Administration; Environmental Protection Agency; and National Aeronautics and Space Administration.

[2] The National Academy of Sciences, chartered by Congress in 1863 to advise the federal government on subjects of science and art, is a quasi-public honorary organization to which scientists are elected annually by vote of the membership. The academy in turn created the National Research Council, which can call upon respected scientists and engineers—who are not necessarily academy members—to serve on voluntary committees.

[3] In addition to research and development, OPA allows the fund to be used for, among other things, oil spill removal costs; payments to federal, state, and Indian tribe trustees to conduct natural resource damage assessments; and payment of claims for certain uncompensated removal costs and damages.

[4] Six agencies reported that they did not conduct oil pollution research: the Army Corps of Engineers, Department of Energy, Federal Emergency Management Agency, Fish and Wildlife Service, Maritime Administration, and National Institute of Standards and Technology.

[5] NRT is the interagency organization responsible for planning and coordinating responses to major discharges of oil or hazardous waste in accordance with the National Oil and Hazardous Substances Pollution Contingency Plan, which is the federal government's blueprint for responding to both oil spills and hazardous substance releases with the purpose of developing a national response capability and promoting overall coordination among the hierarchy of responders and contingency plans. According to EPA officials, the first contingency plan was developed and published in 1968 in response to a massive oil spill from the oil tanker, *Torrey Canyon,* off the coast of England the year before. More than 37 million gallons of crude oil spilled into the water, causing massive environmental damage. To avoid the problems faced by response officials involved in this incident, U.S. officials developed a coordinated approach to cope with potential spills in U.S. waters.

[6] We could not identify the number of projects completed in 2000 because the interagency committee was not required to report on its progress for those 2 years per the Federal Reports Elimination and Sunset Act of 1995, and did not do so. Additionally, we could not identify the number of projects completed in fiscal years 2001 and 2002 because the interagency committee's biennial reports did not include projects; instead, the reports included publications authored by member agencies, and we could not confirm whether individual publications corresponded to a single project.

[7] Nonpetroleum oils include synthetic oils, such as silicone fluids, and seed oils from plants, among other things.

[8] Through the Gulf of Mexico Initiative, NASA conducts applied research and development on weather and climate change to enhance the ecological and economic health of the Gulf of Mexico by using remote sensing, oceanography, coastal processes, signal processing, and mathematical modeling. The initiative was created in 2007 in response to a series of hurricanes in 2005, including hurricanes Dennis, Katrina, Rita, and Wilma, according to NASA officials.

[9] Remote sensing involves gathering data and information about the physical world by detecting and measuring signals composed of radiation, particles and fields emanating from objects located beyond the immediate vicinity of the sensor devices.

[10] According to a BOEMRE report on trends in oil production in the Gulf of Mexico, in 1997 the Gulf of Mexico average annual oil production rates (in thousands of barrels a day (Mbpd)) were 830 for shallow water and 296 for deepwater. In its report, BOEMRE defined shallow water production as production from oil wells in less than 1,000 feet of water and deepwater

production as production from oil wells in greater than 1,000 feet of water. By 2000, the rate was 690 Mbpd for shallow water and 743 Mbpd for deepwater. In 2007, the difference between shallow water and deepwater production had increased, with shallow water production at 381 Mbpd and deepwater production at 895 Mbpd.

End Note for Appendix I

[1] GAO, *Results-Oriented Government: Practices That Can Help Enhance and Sustain Collaboration among Federal Agencies*, GAO-06-15 (Washington, D.C.: Oct. 21, 2005).

INDEX

D

G

H

I

J

L

M

R

S

T

U

V

W